PLASTIC FANTASTIC

How the Biggest Fraud in Physics Shook the Scientific World

Eugenie Samuel Reich

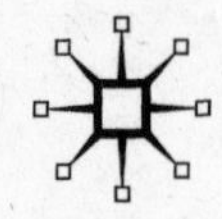

PLASTIC FANTASTIC

First published in 2009 by PALGRAVE MACMILLAN® in the US—a division of St. Martin's Press LLC, 175 Fifth Avenue, New York, NY 10010.

Where this book is distributed in the UK, Europe and the rest of the world, this is by Palgrave Macmillan, a division of Macmillan Publishers Limited, registered in England, company number 785998, of Houndmills, Basingstoke, Hampshire RG21 6XS.

Palgrave Macmillan is the global academic imprint of the above companies and has companies and representatives throughout the world.

ISBN-13: 978–0–230–22467–4
ISBN-10: 0–230–22467–9

Library of Congress Cataloging-in-Publication Data
Reich, Eugenie Samuel.
Plastic fantastic : how the biggest fraud in physics shook the scientific world / Eugenie Samuel Reich.
p. cm.
Includes bibliographical references and index.
ISBN 0–230–22467–9
1. Schön, Jan Hendrik, 1970– 2. Physicists—Germany—Biography.
3. Fraud in science. I. Title.
QC16.S26456R45 2009
530.092—dc22
[B]

2008051801

A catalogue record of the book is available from the British Library.

Design by Letra Libre

First edition: May 2009
10 9 8 7 6 5 4 3 2 1
Printed in the United States of America.

CONTENTS

Introduction 1

1. Into the Woods 11
2. Hendrik 27
3. A Slave to Publication 45
4. Greater Expectations 65
5. Not Ready to Be a Product 85
6. Journals with "Special Status" 105
7. Scientists Astray 129
8. Plastic Fantastic 151
9. The Nanotechnology Department 163
10. The Fraud Taboo 183
11. Game Over 209

Epilogue 235

Notes and Additional References 241

Index 259

INTRODUCTION

The Schön case broke in the news during my first week in a new job. After two years reporting on science for *New Scientist* magazine, I was starting out as a features editor, and might have spent my first days finding writers and commissioning articles. Instead, I found myself immersed in the most riveting piece of technical writing I have ever come across.[1] It was the final investigation report into misconduct in the work of Jan Hendrik Schön, an up-and-coming physics and nanotechnology wunderkind. For scientists, the release of this report in September 2002 marked the end of Schön's story. For me, it was the beginning.

At the age of thirty-one, Schön, a physicist at Bell Laboratories in New Jersey, was emerging with breathtaking speed as a star researcher in physics, materials science, and nanotechnology, a cutting-edge field that involves studying materials on scales as small as an atom. Schön's work was based on his seemingly unrivaled ability to transform the properties of materials by applying an electric field to their surface. He built high-performance transistors, the devices that switch current inside computers, made not from silicon but from plastics and other materials made from carbon. He coaxed materials into superconductors, which have an almost magical ability to conduct electricity without heating up and wasting energy. He described the world's first organic electrical laser, and the first-ever light-emitting transistor. He even claimed to have built the world's smallest transistor by wiring up a single molecule. It was a dazzling nanotechnology triumph.

Schön published his claims in the two most prestigious scientific journals, *Nature* and *Science,* was promoted by Bell Labs, written up in newspapers, and invited to be a guest on National Public Radio, taking calls from enthusiastic members of the public about the state of U.S. science and technology. He won several thousand dollars' worth of prizes, and was offered positions at top research institutions in the United States and Europe. Judging by the number of high-profile articles he published, he was the most productive young scientist in the world.[2]

But the report I was reading that first week, authored by a panel convened by Bell Labs, laid out a series of shocking revelations. Schön's data were fake. His discoveries were lies. Many of his devices had probably never even existed. Even when compared to other major cases of scientific misconduct—from Eric Poehlman, the first U.S. scientist to be jailed for fraud in 2006, to the exposure of South Korean cloning pioneer Woo Suk Hwang—Schön's case was uniquely large in scale. It was unprecedented in terms of the number of discoveries he faked and the number of other scientists he misled or deceived; by my reckoning, scientists in at least a dozen laboratories wasted time and money chasing rainbows because of him. And because of Schön's association with Bell Laboratories, a renowned institution that had shaped the development of modern-day electronics and telecommunications, the scandal raised disturbing questions about how easily the experts we depend on for technological progress can be misled, and what it takes for them to bring their fields of research back on track.

Schön was caught because of the detection of a pattern of duplicated, inconsistently captioned data spread throughout his work. This evidence was collated by two scientists outside Bell Labs, Paul McEuen and Lydia Sohn, who knew that other researchers had failed to replicate Schön's claims in their own laboratories. As McEuen and Sohn informed editors at the major scientific journals, Bell Labs managers called in a panel of investigators who found Schön guilty on sixteen charges of misconduct, including intentional fabrication of data. Schön greeted the investigation report with a statement in which he said he regretted making mistakes in his research but insisted that he had really seen all the physical effects he had reported.

In interviews at the time, outside scientists demanded to know how things had been able to get so far. How had one person—the investigators cleared everyone else of fraud—succeeded in convincing managers at a top U.S. research institution to back and promote such grossly fabricated claims? Why had so many scientists, including Nobel Prize–winners and others respected for scientific judgment, recommended Schön for prizes and top jobs? Why did journals trusted to deliver news of scientific developments publish his papers? As the physicist Paul Grant wrote at the time, the system seemed to work, because the fraud came to light, but the scale of the case begged the question why it hadn't worked sooner.[3]

Reporters like me tried to understand these issues when the story broke, but at the time it was impossible to resolve them. People with first-hand knowledge of the fraud were reluctant to talk to us, or were prevented from doing so by the company that owned Bell Labs, Lucent Technologies. Many scientists also seemed caught up in an angry blame game, and were unable to share judgments about whether this could have happened in their own laboratories, and if not, why not. The question of what would make a smart young scientist take such dramatic risks continued to burn. Was Schön one of a kind or was he the tip of the iceberg? Why did he do it? Why, given the much-celebrated self-correcting nature of science, didn't he realize he would get caught? On the day of the report's release, Schön was fired and fled the United States to an unknown location. Answers to these questions seemed to vanish with him.

Two years later, a colleague forwarded to me a press release that reminded me of Hendrik Schön. It announced the invention of transistors similar to some of those that Schön had claimed to make. This time, the devices were real, built by researchers at Rutgers University, a few dozen miles from Schön's former lab at Bell Laboratories' site in New Jersey. The Rutgers group had used techniques different from Schön's and obtained results that were in many ways more modest, but the same physical principle of turning carbon-based materials into electrical switches was demonstrated by their experiments.[4]

The investigators of Schön had considered this possibility, writing in their report that the finding of scientific misconduct against Schön would remain valid even if the science of Schön's claims was validated in the future. Even so, it was interesting to think about what might have happened if this research, or other work similar to Schön's, had been completed sooner. Schön might have earned the credit for being one of the first to jump into a novel area, while the scientists who did the work to test his claims appeared to come in second. Paul McEuen and Lydia Sohn might never have searched for evidence of fraud, and Schön might have gotten away with it. Suddenly, the self-correcting nature of science looked as if it could slice both ways.

Hooked by this idea, I couldn't help but wonder whether it had occurred to Schön too. Had Schön been banking on the possibility that his false but plausible scientific claims might one day be validated through the honest work of others? At the same time, the appearance of similarity between Schön's work and other genuine results helped to explain why other scientists had been so willing to believe Schön in the first place. Schön had apparently imitated the outline of real scientific breakthroughs well enough that his data seemed both groundbreaking and plausible at the same time.

Following up on this idea, I began a series of interviews to try to understand the Schön case once again. With two years having passed, I found scientists easier to engage than they had been, and I left my job at *New Scientist* to follow the story. Over the next three years, I carried out email, phone, or in-person interviews with 125 scientists and journal editors who had interacted with Schön (thirteen declined interviews, nine did not get back to me), and pieced together the human story behind the findings of fraud released by Bell Labs managers. Those who responded included some of Schön's former collaborators, managers, and colleagues inside and outside Bell Labs, many of whom had kept documents, data, unpublished manuscripts, and emails from the time. I also had contact with Schön's former university, the University of Konstanz, and with the German Research Foundation, which had funded some of his work. I obtained further documents from the U.S. Patent Office, and I used the Freedom of Information Act to obtain records from the U.S. National Science Foundation. I also saw a large amount of contemporaneous email from inside and outside Bell Laboratories.

From other scientists I learned an enormous amount about the ideas, experimental suggestions, and feedback provided to Schön during his time in science. Comparing this to the claims he made, I found a pattern of compelling resonance. Schön was apparently working furiously to integrate the research ideas of the scientific community around him into his publications. No wonder, then, that scientists were so thrilled by his papers. Schön had turned their best ideas into fabricated data that were bound to seem appealing. This helped to explain both why his claims got a good reception and why those claims had something in common with results later achieved in reality by other scientists. Schön hadn't guessed the onward course of science as much as listened to colleagues who had.

Looking back in history, the idea that scientific fraudsters transform human ambitions and preferences into data has a precedent. In 1830, the English mathematician Charles Babbage wrote an essay called "Reflections on the Decline of Science in England, and on Some of Its Causes" that is often cited as the classic introduction to scientific fraud.[5] Babbage described a truth-seeking scientist as a careful observer who goes to great lengths to try to prevent his bias from influencing the facts he reports. A scientific fraudster does the opposite, consciously allowing preferences to interfere with the reported observations, Babbage wrote. Babbage imagined several different kinds of fraudster but admitted there was one type that he could not understand: a forger, who not only manipulates or improves data, but makes things up. In Babbage's vocabulary, Hendrik Schön would be a forger. There have been forgeries in science before—the most famous example is Piltdown man, a forged fossil that surfaced on Piltdown Common in England in 1912—but Hendrik Schön's case marked a watershed in scientific forgery because of the breathtaking extent to which he had repeatedly pulled high-profile electronics breakthroughs out of thin air.

My starting point in understanding a question that Babbage could not—what drives a forger—was to look at Schön's data. As Babbage wrote, forgers craft their data to suit their view of a situation. Misleading as this is for other scientists, it also provides an opportunity because it means that fraudulent data hold a tremendous amount of information about what a forger was thinking. I think that this is one reason why I found the official investigation report into Schön's case so riveting. In the investigators' dissection of Schön's fraudulent data was the occasional fascinating glimpse into what Schön must have been thinking as he faked it. The idea that fraudulent data could betray a state of mind is familiar to whistleblowers and many investigators of irregular science, who have looked through irregular data to discern tell-tale signs of intentionality. Intent, or motive, is the key difference between errors, which scientists consider an inevitable part of cutting-edge science, and fraud or forgery, which is considered an aberration.

Not content with looking at Schön's data, I also contacted Schön. This turned into something of an adventure. After making efforts to track Schön down in Europe, I received email from a person who seemed to be writing under a false identity. The pseudonym used the email address plastic_fantastic@hotmail.com. The phrase; "Plastic Fantastic" had been used in the media to describe Schön's measurement of superconductivity in polythiophene, a plastic material. I engaged strongly with the correspondent, figuring that if I could win his trust, it might lead to an interview with Schön. That did not happen, and I eventually decided not to quote or rely on the strange emails. When I called Schön on the telephone, he told me that he was not interested in talking. His official position remains that he believed he had seen the physical effects that he reported, at the time that he reported them. Nevertheless, partly by obtaining from other scientists emails that he had sent and received during his time in science, I formed a picture that went some way beyond this. Talking with Schön might not be the best way to understand him, I realized; he was so responsive that any interviewer would risk eliciting the statements that he or she wanted to hear. I came to understand him mostly through his fabrications, charting the way that this obliging and literal-minded young man developed into one of the most brazen fraudsters of our time.

To place Schön in historical context, I took occasional detours from his case. I looked at the case for fraud in some of the work of Isaac Newton, whose psychology has been the subject of extensive work by historians. I looked forward to the recent cases of Eric Poehlman and Woo Suk Hwang. I looked at several other cases, old and modern. I learned that the motive and modus operandi of fraud could often be understood in the context of the science of its day. This is not only because fraudsters are very responsive to the incentives and pressures around them. It is because any scientific community has ways of checking the validity of scientific claims, and the fraudsters know what the checks are.[6]

In 2003 a senior manager who oversaw Schön's career at Bell Laboratories, Cherry Murray, together with a Bell Labs press officer, Saswato Das, wrote a commentary for the journal *Nature Materials*. On the one hand, Murray and Das expressed the view that all was well because the fraud had come to light. "The beauty of science is that it is self-correcting. The mills of science grind slowly, yet they grind exceedingly small," they wrote.[7] On the other hand, Murray and Das claimed that nothing could have been done to prevent Schön's misconduct because "if someone is determined to be unethical, it is not easy to detect," and because it would be unfair to question the integrity of others without proof. According to this view, it was inevitable that Schön would succeed in perpetrating his fraud under the eyes of Bell Labs managers and inevitable that he would get found out.

Perhaps unsurprisingly, my findings are different. But my story also revolves around the idea that science is self-correcting. The essence of this idea is that even if individual researchers make mistakes or commit fraud, other researchers, working in different laboratories or institutions, are bound to set things straight again. It is frequently advanced by people who work with a big-picture view of the scientific community, such as journal editors, who are responsible for evaluating new scientific claims, and scientific managers, who oversee dozens of lab researchers, but mostly no longer collect data themselves. It is also a very influential idea, often invoked in support of the view that science should be left to regulate itself, with minimal interference from institutional officials or government agencies.

Schön's story represents a challenge because Schön sailed through many of the stages in science's self-correction process and was caught, in the end, only in a very peculiar way. In the mid-1990s, Schön started out as one graduate student of many in the laboratory of the physics professor Ernst Bucher at the University of

Konstanz in Germany. (The University of Konstanz is in Constance on the Swiss-German border. I will use the town's German name, Konstanz, to be consistent with the name of the university.) In Konstanz, Schön was considered a solid student, although, as I discovered, he was quietly already beginning to develop problems in handling scientific data. But crucially, it was not until Schön moved to Bell Labs in the United States that his fraudulent claims escalated. At Bell Labs, Schön worked with a supervisor and manager, Bertram Batlogg, two close colleagues, Christian Kloc and Zhenan Bao, and over a dozen other collaborators. His work came under the scrutiny of managers, including Federico Capasso, who had responsibility for managing around one hundred researchers; John Rogers, who became Schön's department head after Batlogg left; and many staff members. From there, Schön's data passed through review at the scientific journals and were well received in the scientific community, even after it became known that his findings were hard to replicate. My story follows Schön and his claims through these stages, watching out for points in the story where the self-correcting nature of science might kick-in.

The investigators of Schön had laid some of the groundwork. In their final report, they suggested that Schön's collaborators might ideally have acted as the first line of defense against fraud, but nevertheless they cleared all of them conclusively of misconduct. They added that questions had arisen about the way in which Schön's managers, and the scientific journals, had performed. Given also that Schön had been able to work without keeping proper records of his experiments, the investigators suggested that his case raised questions about the way that scientists' record-keeping practices have evolved in the computer age.

Also clear was that Schön had joined Bell Labs at a very difficult time in its history. Owned by the telecom company Lucent Technologies, Bell Labs was caught up in the excitement of the Internet boom and plunged into turmoil by the falling revenues and collapsing share prices of the Internet bust. Researchers at Bell Labs were exposed to the suggestion that they might publish high-profile research papers, and in the late 1990s the oversight of research began to ease off. This phenomenon is not unique to Bell Labs, as almost all scientists, including those at universities, are working with their next grant application or major publication in mind, and it is not unheard of for researchers working on a project that is under threat to promote preliminary data more than they otherwise might. Add to such a situation a person who is willing to fake data to fill the desires of those around him, and add to that the natural excitement that scientists feel about any new result, and it becomes easier to understand how fraudulent claims can gather momentum.

But this was not the whole story. Repeatedly, scientists I interviewed told me that they had known that aspects of Schön's work were unconfirmed, or even that they knew of specific problems, and yet they still failed to detect and stop his fraud. When I asked why, interviewees often invoked the idea that science relies on trust. It had been natural to question the way that Schön had interpreted his data, they suggested, but not to assume that he was lying. When I asked whether placing so much trust in the statements of an individual researcher was not too much of a risk, I sometimes encountered agreement, but was frequently told that this was not the case. The idea kept coming up that science is self-correcting. If a scientist lies he or she will be found out eventually. Pouring too much cold water on new results would be bad for creativity and would stifle the scientific process. Ironically, at numerous points in this story, scientists' implicit assumption that future research would resolve outstanding issues operated as part of the rationalization for allowing Schön to keep going.

Of course, the idea that science is self-correcting is not only subscribed to by scientists. Many non scientists believe in it too, and with good reason. Over the past two centuries, science has delivered tremendous technological and medical advances. We feel wealthier and healthier, and we have a better understanding of our place in the universe than previous generations did, in large part because of science. We all tend to trust airplanes, cars, computers, hospitals, doctors, and technical advice from qualified experts.

But while it is clear that science and technology have been very successful, it is less clear to most of us why. Why do some scientists feel driven to fake results? How do others know which claims are true and which are false? What circumstances enable experts to establish the truth about new claims now, rather than next year, or maybe never? Are there ways to give novel results a fair hearing without a whole institution or community going astray? How much in science should be questioned, and how much should be taken on trust?

Without the answers to such questions, it seems like little more than blind faith to insist that all activity carried out in the name of science will always be self-correcting.[8] Science is a human activity, after all. The scientific method does not come naturally to most people but has to be learned over several years of training. Many of the research institutions in which science is taught have been founded fairly recently; Bell Laboratories was only seventy-seven years old at the time that the Schön case made the news. So what do you expect? Is the story of an extensive case of fraud that was eventually brought to light a story about how science succeeds in self-correcting? Or is a story about how it fails?

1

INTO THE WOODS

Bertram Batlogg wasn't feeling the slightest bit surprised. He was captured on camera at the back of the stage, his hands clasped in applause, with a slight smile on his face as photographers took pictures of the executive vice president for research, Arun Netravali, embracing the newly crowned Nobel Laureate, Horst Störmer. It was wonderful to hear that the most prestigious prize in physics had been awarded to a close colleague, Batlogg later told a press officer.[1] It was also news that Batlogg had been expecting after more than a decade as a manager at Bell Labs.

It was October 1998, and Murray Hill still felt the same. America's most illustrious industrial lab was tangled in feeder roads for Interstate 78, a forty-five-minute drive from New York City. The New Jersey estate held a complex of pale orange buildings, topped off with an upside-down V of copper-oxide green rooftops that looked like wings poised to spread. Behind the buildings, the sharply sloping woodland blocked out sight of the highway.

Bell Labs research took place in the shadow of a remarkable history. In 1947, researchers at Murray Hill invented the transistor, laying the foundation for the development of the silicon chip and the computer industry. Bell Labs researchers did much of the groundwork toward the invention of the first laser, and they were the first to use radio telescopes to detect the echo of the Big Bang. With researchers in fields from astrophysics to the manufacture of optical fibers, Bell Labs earned a reputation for ambitious but reliable research that had repeatedly led to technological progress.

For over half a century, Bell Labs had been owned by the telephone monopoly AT&T and had plenty of money to spend on science. But in 1984, the monopoly broke up, and after 1989, managers encouraged Bell Lab researchers to focus on research with commercial applications. Disenchanted, top scientists began to leave for universities and were mostly not replaced by new recruits. In 1995, ownership of Bell Labs was transferred to the newly formed company Lucent Technologies. It seemed as if the prospects for science were about to get even worse.

But that didn't happen straightaway. What came next was a trend-defying scientific revival. The late 1990s were the swinging years of the Internet boom, and as a telecom company Lucent flourished, and so did its scientists. After several years of decline, the number of discoveries published by researchers at Murray Hill began to rally. The rate of patent filings on new inventions was on the increase too. Then, in October 1998, the announcement came that Bell Labs' Horst Störmer had won the Nobel Prize for Physics. As the good news spread, Bertram Batlogg was one of hundreds of scientists to leave his office and make his way down the steel-walled utilitarian passages of the research buildings into Murray Hill's vast cathedral-like cafeteria to celebrate a bit of quantum physics that had never made the company's shareholders so much as a dime.

In 1982, Störmer and Daniel Tsui of Bell Labs had made the first measurements hinting that electrons, the negatively charged particles that transport electricity on computer chips, can be made to split apart in a strong magnetic field. The finding had been unexpected because electrons had previously been considered indivisible. But it had been explained in detail as a counterintuitive consequence of quantum mechanics, the theory that describes the way particles behave under the influence of the universe's fundamental forces. The theoretical physicist Robert Laughlin shared the 1998 Nobel Prize with Störmer and Tsui for the discovery that had become known as "the fractional quantum Hall effect." (The nineteenth-century American physicist Edwin Hall had discovered the non quantum Hall effect.)

By 1998, Horst Störmer had moved to Columbia University in New York, but he was still affiliated with Bell Labs. He left Manhattan on the day of the Nobel Prize announcement and traveled to New Jersey to share the glory. The Murray Hill cafeteria was decked with balloons. There was free cake. It was the third time in three years that research associated in some way with Bell Labs had been recognized by the Nobel Prize Foundation, and people joked to one another: "Do we win a Nobel Prize every year?" But it wasn't only on the inside that the prospects for science at Bell Labs seemed bright. Out in California, even Laughlin, Störmer's cowinner and a ready critic of poor science management, took a moment out of his Nobel Prize press conference to praise Lucent for putting science back together.

To Bertram Batlogg, who had known Störmer for years, it seemed, if anything, surprising that the Nobel Prize hadn't been awarded for this work sooner. But the celebration was an opportunity to look back on the good old days of science under AT&T, and to feel optimistic about the future. Lucent Technologies' web site[2] quoted Batlogg describing staff scientists as "elated" and hopeful that the recognition would attract the best scientists to Murray Hill.

Among Batlogg's interests at the time was a research program to explore the industrial potential of plastics. He had already hired Jan Hendrik Schön, a postdoctoral researcher from a lab in Germany, to join the program. Over the next three and a half years, Schön went on to make a series of discovery claims that turned out to include some of the most outrageous lies ever to be exposed in modern physics. After Schön's disgrace, Batlogg and other managers, steeped as they were in the Bell Labs tradition of research excellence, emphatically insisted that Schön's fraud was something they couldn't possibly have seen coming.

"IT WAS NATURAL FOR A MANAGER TO ASK"

The chain of events that led Batlogg to recruit Schön had been set in motion three years earlier. It started, Batlogg later recalled, during a performance review. Batlogg was the head of Bell Labs' Department of Materials Physics. He managed around a dozen staff researchers, known as a "Member of Technical Staff" or "MTS." The MTSs ran their own laboratory, and every year filled out a form listing scientific contributions to be assessed by managers. In the weeks leading up to these reviews, Batlogg often noticed technical memoranda piling up on his desk, and he warmed to his staff's burst of productivity. He tried to speak up for them, but department heads graded in a group so that nobody could inflate the grades of his or her own people. The reviews turned into debates about the impact of science in one department versus the impact of science in another, and Batlogg, an appealing sandy-haired manager, had no trouble holding his own. He was a charming, ambitious, and also very competitive person who mostly enjoyed the back and forth of heated discussions and regarded them as a way of getting to the bottom of the most interesting scientific questions.

In 1995, one of these questions arose from the work of a team of Bell Labs researchers who had been making transistors out of plastic. Transistors are the devices that switch current inside computers. A modern computer contains as many as a billion transistors, wired together on silicon chips in circuits that process information. At first blush, the replacement of silicon with plastic inside transistors should sound revolutionary. Plastics are mostly thought of as insulators, materials that are electrically dead and unable to conduct current. Silicon and other materials used in transistors are semiconductors, materials that can change their properties from insulating to conducting under the right conditions. But scientists have known since at least the 1970s that some plastics are semiconductors and, in principle, could be used to make transistors too. The prototype plastic transistors being made at Bell Labs in the mid-1990s conducted current hundreds of thousands of times more slowly than commercial silicon transistors, but research had progressed to the point where, Batlogg gathered, some companies might be ready to market the first plastic computer chips within five or ten years. Plastic electronics were expected to be better than silicon for some applications, such as light, flexible radio identification tags that are used to track goods in chain stores like Walmart or the Gap, or computer screens so thin and bendable that they might one day be sold as "electronic paper." Plastic chips might also be more easily woven into clothing or accessories. This quest for a versatile, practical electrical plastic eventually provided the commercial backdrop to several of Jan Hendrik Schön's fraudulent discovery claims.

But the relationship between the possible product applications and science was not simple. One of the Bell Labs' MTSs working with plastics, Ananth Dodabalapur, and his colleagues were claiming that the flow of current through their devices was as fast as would ever be possible for plastic electronics.[3] But the claim was controversial, and became even more so when researchers at Pennsylvania State University reported making devices in which electrical charge seemed to move faster.[4] To Batlogg this raised an interesting question. If the current prototypes weren't reaching the limits of plastic electronics, what would?

Also present at performance reviews was Bob Laudise, a senior manager nearing retirement who had provided some of the impetus for the company's effort to develop plastic electronics. Laudise had a background in crystal growth, the skill of transforming chemicals into crystals. Knowing that plastics are made out of rings or chains of carbon atoms that are arranged in a chaotic tangle, Laudise and Batlogg discussed the possibility that the chaotic internal structure was slowing the movement of electrons through the materials. Could the molecules in plastics be rearranged into a more orderly way? The ideal form would be an organic crystal, in which the irregular organic molecules would be arrayed in a regular way. To visualize how the interior of an organic crystal might look, think of the regular way that irregularly shaped sardines can be stacked neatly in a tin. Batlogg was struck by the idea that electrical measurements on organic crystals might be able to reveal the ultimate limits of plastic electronics because the effect of chaos and disorder would be reduced to a minimum. And once those limits had been reached, it would be possible to address a host of new scientific questions. For example, did electricity flow through plastic and organic materials in the same way as it did through silicon and other semiconductors? Could the new materials reveal anything about the quintessential nature of matter and electricity? Might previously unknown quantum effects be discovered?

Several of these questions were scientifically interesting, but the answers weren't expected to lead directly to new plastic electronics. As often happens in science, the managers could not be sure how having a deep understanding of organic crystals would bring about technological progress. But they still felt confident that it would. At Bell Labs, founded in 1925 as the innovation powerhouse of AT&T, it was traditional to argue that there was a virtuous spiral between science and technology. To explain this idea, attributed to the Dutch physicist Hendrik Casimir, the managers liked to give historical examples. The first transistor had been created at Bell Labs in 1947 following fundamental research on the properties of semiconductors. The device was memorialized by the display of a replica in the lobby entrance to Murray Hill. Nearby was a bust of Alexander Graham Bell, the nineteenth-century inventor of the telephone. In one of his lectures, Bell had coined a phrase that became Bell Labs' motto: "Leave the beaten track occasionally and dive into the woods. Every time you do so you will be certain to find something that you have never seen before."[5] To be truly innovative, an industrial lab needed not only product development, but also a culture of free-thinking scientific exploration. Scientific managers at Bell Labs were always looking for research projects that appeared to embody this idea.

Organic crystals were a good example. Once plastic electronics had been identified as a possible future technology for Lucent, the managers wanted to lead the world in understanding the fundamental scientific principles that governed the electrical behavior of the materials. Then engineers working on product development would be well placed to make rational decisions, rather than having to grope in the dark. "The idea was that you couldn't develop something as a technology without understanding fundamentals, that this was putting the cart before the horse," said Peter Littlewood, who followed the discussions as head of the Department of Theoretical Physics Research. Often, managers taking this view would blur the line between science and technology, so that when I asked Batlogg whether the research program on organic crystals was fundamental science or applied science that was relevant to technology he replied that "the question makes no sense, because there is no difference between those things."

For scientists in the Physical Research Lab, an organization of about one hundred researchers that included Batlogg's department, the maintenance of this blurred line between science and technology was a question of survival. Researchers in the Physical Research Lab did more fundamental science than anyone else at Bell Labs. If Lucent decided to cut back on science, this lab would be the first to face the axe. As for Bertram Batlogg, he was best known not for his contributions to product development but for his scientific contributions. According to one widely circulated list that ranked physicists by citation (the number of times their work was mentioned by others), Batlogg was fourth, with two Bell Labs colleagues second and third, in the entire world.[6] In his heart, Batlogg was also far more of a scientist than an engineer. For example, when a dent appeared in the family car, Batlogg apparently tried to save a few bucks by fixing it at home with a crowbar, and, the story goes, the crowbar snapped. An engineer might have been embarrassed, but as a scientist who appreciated the fundamental properties of materials, Batlogg's reaction was to feel impressed by the strength of the steel in the flanks of his American car.

Traditionally, Bell Labs had a place for scientists who were curious about the fundamental properties of materials, but who were not as practical as their engineering colleagues. And anyone interested in the intrinsic nature of matter would have to be interested in the organic crystals envisioned by Bob Laudise. Five years on, when the Bell Labs research program appeared to be developing into a huge success, Batlogg gave a seminar in which he looked back at the early days and explained that the more orderly the molecules in plastics, the faster the charge might move, and so the better the potential performance of plastic electronics would be. Although Batlogg had not known what the future held for the new technology, he had been confident that, by making crystals, Bell Labs could stay a step ahead of competitors in exploring where the new technology would lead. "It was natural for a manager to ask, 'Where does it end?'" he said.[7]

BERTRAM BATLOGG AND HIGH TEMPERATURE SUPERCONDUCTIVITY

Bertram Batlogg was known for his work on superconductors, materials that are able to conduct electricity without putting up resistance.

In contrast to semiconductors, which are useful for computing because they can change their electrical character from insulating to conducting, thus switching current on and off and so processing information, the most obvious potential application of superconductors is to transport large amounts of electrical power, for example in national power grids. But the Dutch physicist Heike Kamerlingh Onnes first detected superconductivity in mercury in 1911 at a temperature only a few degrees above absolute zero. It seemed that any money saved by using superconductors in practice would be outweighed by the expense of making the materials cold enough to work.

In 1986 the outlook changed. Georg Bednorz and Alexander Müller at the IBM Zurich Research Laboratory reported on the discovery of a cuprate, a compound containing layers of copper and oxygen atoms that became a superconductor at the relatively higher temperature of thirty-five degrees above absolute zero. This was high enough that it began to seem practical to try to develop a superconductor that would work at practically attainable temperatures, including perhaps even room temperature—the temperature at which we live.

Batlogg's group at Bell Labs (as well as many other researchers) followed up rapidly on the Swiss claim by measuring and characterizing the properties of other cuprates. They came across compounds that turned superconducting above seventy-seven degrees, a temperature that, while still a long way from room temperature, could be reached using liquid nitrogen, an affordable coolant. In 1987, the Swiss team won the Nobel Prize in recognition of their role in opening up a new field of research.

Batlogg was promoted to be the head of the Materials Physics Research Department at Bell Labs shortly before the field of high-temperature superconductivity or "high-Tc" took off. Alongside his departmental responsibilities, he continued to do science, working closely with two American colleagues in other departments. Bob Cava, a chemist, made samples of the new superconductors, while Bruce Van Dover, a physicist, made electrical measurements. Batlogg's laboratory was responsible for measurements of the way in which the new samples responded to magnetic fields, a type of measurement in which he was an expert. The team worked fast, by trusting each others' expertise. "If Cava told me something was 10% strontium, or Batlogg told me it expelled 40% of the magnetic field, those were facts," Van Dover explained. The relationships were also candid. Batlogg was born in Austria and came to Bell Labs after completing his PhD in 1979 at the Swiss Federal Institute of Technology in Zurich. When he bridled at the way that Americans sometimes abbreviated his first name from the Austrian form, "Bertram," to the American form, "Bert," adding that "Bert" sounded like a character from *Sesame Street,* Bob Cava responded by signing his own name "Ernie" in a memo. Batlogg was privately amused.

The rising profile of high-Tc meant that the Bell Labs effort needed a spokesman, and Batlogg took on this role. He argued Bell Labs' case for priority on important discoveries and captivated audiences at international conferences with a gift for expressing new information in such a way that other scientists would understand right away. In one presentation, he plotted all the known superconducting materials as dots on a chart, with the position of each dot determined by two properties—the temperature at which the material became superconducting and the number of electrons available to conduct current in the material. Dots representing old superconductors clustered together while high-Tc materials were off the main trend, revealing to the audience in what ways new theory and experiments were needed. "Off the beaten track," Batlogg liked to say, purposefully echoing the Bell Labs motto. Studying materials that lay on the known trend would have been "stamp-collecting," a derogatory formulation that Batlogg used for science that he found repetitive or boring.

But the group owed its success in large part to caution. Van Dover recalled how, in the heady atmosphere of high-Tc, their group at one point wrote a very exciting paper that claimed the discovery of a material that was superconducting at a record high temperature, 240 degrees above absolute zero. "Over the weekend we slept on it," said Van Dover, "and I think Monday morning we looked at each other and Batlogg said, I think it was Batlogg, 'This doesn't meet the Bell Labs standard' and we agreed." They called off the submission, which made them feel lousy, because of the possibility of another group beating them to the priority claim. But the decision turned out to be right because further experiments did not bear out the result. Other scientists came to realize that when Batlogg's group made a claim public, they were quite likely to be right.

But the good times didn't last for high-Tc. Over time, the lab work became more technical. The titles of talks got longer. No one discovered a room-temperature superconductor, and theoretical physicists struggled to come up with a theory that could explain the effect. One ambitious young physicist who had thought of applying to Batlogg for a job in the mid-1990s told me he decided against it because he felt that with high-Tc having gone off the boil, Batlogg, while very influential, "wasn't as hot as he used to be." Batlogg disputed this suggestion when I put it to him, but he did not dispute that he had been on the lookout for promising new areas of research that might explode to fill the void left by the domestication of high-Tc.

Organic electronics was one of these areas. High-Tc had shown that one way to open up a new research area was to study a class of materials with an unusual structure, as the Swiss had done by choosing cuprates in 1985. A decade later, Batlogg conceived of a research program that involved exploring many different properties of organic crystals. It even seemed possible that something as profound as high-temperature superconductivity, but of a very different nature and with different applications, might be discovered. And then how would the research proceed? "We give samples out to other groups to repeat measurements, or do other measurements, but they would have to cite our work, because we were first," said a former intern recruited to the effort. So although organic crystals were expected to be relevant to plastic electronics, Batlogg did not expect to measure the reward for success in commercial terms as much as in terms of the recognition that could come from being the one to open up a new field of research. In the words of Alexander Graham Bell, he was diving into the woods.

Things got off to a tricky start. In 1995, Bell Labs researchers were working with alpha-sexithiophene, a molecule that was known to conduct electricity better than many other organic materials. Alpha-sexithiophene had a complex structure, consisting of six rings of carbon and sulfur atoms linked together. At the time, so few labs were working with the chemical that Bob Laudise could not order the quantity he needed from a standard chemistry catalog and had to seek out a specialist supplier. In his first attempts to make crystals, Laudise also found that the unwieldy, fragile molecules fell apart or ended up as plastic films rather than arranging themselves in an orderly way.

He and Batlogg decided to take a different tack. They knew of a lab in Germany that employed a chemist, Christian Kloc, with an apparent knack for figuring out how to crystallize temperamental substances. To make progress in building up a Bell Labs effort on organic crystals, the managers needed someone like Christian Kloc on board.

THE CRYSTAL GROWER

Crystal growers sometimes seem like practitioners of a magical art. Armed with glass flasks and tubes of all shapes and sizes, bubbling solutions, Bunsen burners, vacuum pumps or canisters of gas, crystal growers aim to hit on just the right recipe to transform a starting chemical into a perfect crystalline form. As in baking an elaborate cake, small variations, such as fluctuations in temperature or contamination of the equipment, can change the outcome from success to failure, which in crystal growing is the difference between a large, shiny crystal and a horrible blob of sludge.

Crystal growth is often spoken of with mystique by other physical scientists, and in some labs I visited researchers followed recipes developed by professional crystal growers while admitting that they didn't know why the recipes worked. In the same way, crystal growers often said that they didn't follow what happened to their crystals or recipes after their work was done.

Christian Kloc was accustomed to doing his own thing by personality as well as profession. Born in Communist Poland, he trained and worked there as a chemist until 1985, when he, his wife, and his small son left their apartment, let it be known that they were going on a vacation, and drove west. Once over the border into West Germany, they defected, abandoning their apartment and the belongings they had left in Poland. With a modest approach to life, Kloc later said the only thing he minded having left behind was a stamp collection that he had been keeping up to date every year from the age of four. In Germany, he landed a job with Ernst Bucher, a professor of physics at the University of Konstanz in the south of the country. He worked for Bucher for over a decade, building a reputation for figuring out innovative methods for crystallizing new materials. He was not personally ambitious, however, and was still on the same contract in 1996, when he received a call from Bob Laudise about organic crystals.

It was agreed that Bell Labs would fly Kloc to the French National Center for Scientific Research in Paris, where Gilles Horowitz and collaborators had reported some success making crystals of alpha-sexithiophene,[8] the chemical that Laudise was having trouble with, and that Kloc would write a report on the possibilities of starting organic crystal growth at Bell Labs. During his visit to Paris, Kloc found that the French did not seem to have any specialized equipment, suggesting, he felt, that Bell Labs wouldn't need any either. This meant that the growth of organic crystals could become a much bigger field in the future, as the Bell Labs managers hoped. Kloc also saw that the French seemed to be using as much as half a gram of starting material when they grew their crystals, which to Kloc was bound to increase the risk of introducing impurities. That could explain why the French reported making crystals that were reasonably high in quality, but not so good that the electrical charges in them were entirely free to move without being hindered.

Could Bell Labs do better? There, the first transistor had been invented in 1947 by researchers working with germanium. But in the end, silicon, not germanium, became the staple material for the modern computer industry. So it was well known that the success or failure of electronics might depend crucially on the choice of material. Kloc understood that his job would be to find a general method that would allow Bell Labs to crystallize many more organic materials than alpha-sexithiophene.

Kloc grew his first organic crystals in a vertical glass tube held by a couple of bench clamps in Ernst Bucher's lab. At one end was a small electrical heater that warmed and vaporized a tiny amount of organic material placed at the bottom, and at the other end was a cool region in which molecules condensed from vapor to form crystals. He rejected the idea of trying to evacuate all the air from the flasks, as he figured that the French group had not done this. Instead, he allowed pure, clean argon or hydrogen gas to flow all the way through the apparatus, later finding that a horizontal tube worked better than a vertical one for this. Peter Simpkins, an MTS with expertise in the way that molten glass flows during the manufacture of optical fibers, had suggested that if Kloc got the rate of gas flow right, the bulky organic molecules would have time to rotate gently into the right orientation to attach to the growing crystal. It worked, and long orange-yellow translucent crystals of alpha-sexithiophene formed as oddly orientated flakes growing out every which way from the walls of one flask. Kloc broke the flask open and measured the crystals. They were many millimeters in size, large enough for physicists to make electrical measurements on them.[9] Kloc informed Bell Labs of the success by fax. Laudise faxed back "GREAT."

Kloc was appointed as an MTS at Bell Labs and given his own lab in Building One, one of the research buildings at Murray Hill. He adapted his apparatus to make crystals from many more organic materials than alpha-sexithiophene. He transformed pentacene, molecules of which are made from five rings of the aromatic chemical benzene joined together, into gray crystals. He made tetracene, molecules containing four benzene rings, into orange crystals. Compared to dull, black silicone, many of Kloc's crystals looked like brightly colored gemstones.

Given his own lab and an industry salary, Kloc said he began to understand his own worth for the first time. "At Bell Labs, it doesn't matter who you are, or where you come from, only what you can do," he told me when I visited his lab there in 2005. What he could do was to find practical, table-top ways to grow crystals of new and novel materials. "My skill wasn't about getting things perfect," he explained, "it was about doing things for the first time." He focused on growing and improving his crystals, was reluctant to interfere with others, and always maintained a sunny and positive outlook on the way things seemed to be going.

THE FIELD-EFFECT TRANSISTOR

Modern physics is inseparable from the theory of quantum mechanics, developed in the 1920s to describe the way particles behave. In the theory, physical quantities such as light, sound, and electricity are described both as waves shaped by the universe's fundamental forces, and as individual particles or "quanta." The theory has been used to predict and explain diverse natural phenomena—from the rate at which objects lose heat to the radiation emitted by dying stars. But its equations are not always easy to apply in practical situations, especially when many interacting particles and forces are involved.

Against this background, laboratory experiments on solid materials have enabled physicists to explore more of the implications of quantum mechanics than would be possible otherwise. The Nobel Prize–winning discovery of the fractional quantum Hall effect is one example. In this effect, electrons, which are usually thought of as indivisible particles, respond to the presence of a strong magnetic field by merging together into a kind of fluid. That fluid then splits apart into new particles that are able to carry charges that are one third of that on an electron, making it appear as if the electron has split apart. This was an unexpected finding that no one had calculated prior to the first laboratory experiments. Robert Laughlin, the theoretical physicist who explained it, has since argued that some laboratory experiments have a very deep cosmological significance, revealing phenomena that seem almost unimaginable until they are measured.[10]

The fractional quantum Hall measurements were made in "flatland"—all the electrical charges in the experiment were confined to the surface of a gallium arsenide crystal. As it turns out, other important electrical effects also involve the movement of charge on surfaces.

For example, computer chips process information via the flow of charge on the surface of crystals of silicon or other semiconductors. And high-temperature superconductivity arises from the way in which electrons interact when confined to the planes of copper and oxygen atoms that are found in cuprate materials.

Physicists' imagination about what it might be possible to measure on surfaces also turned out to be a crucial ingredient in Hendrik Schön's historic case of research fraud. In the mid-1990s, the plastic transistors made at Bell Labs were field-effect transistors, which are, in principle, ideal devices for measuring the properties of surfaces. In a field-effect transistor, two metal contacts are attached to the surface of a semiconductor. Typically there are two contacts, known as the source and drain electrodes, which supply and remove electrical charge. Then a third electrode, called the gate electrode, is placed close to the surface of the semiconductor, but without touching it. When a voltage is applied to the gate, an electric field is cast onto the semiconductor, changing its properties and so altering the rate at which charges are able to flow between the source and drain. To ensure that the gate does not touch the semiconductor, which would cause a short circuit, the surface is protected from the gate electrode by a thin layer of insulating material, the gate insulator. Many interesting and profound quantum effects have been detected on this protected surface by physicists making measurements of the current or voltage linked to the electrodes.

To managers at Bell Labs, the organic crystals grown by Kloc seemed as if they might be good candidates for field-effect transistor experiments. The crystals were structured with the organic molecules interlocking to form planes, with the planes stacked on top of each other. If the crystals were cleaved (divided) between two planes, each side would provide a clean surface for field-effect transistor measurements. That was the idea, but the crucial test in science is experiment: could anyone get such transistors to work? Batlogg's background was in measuring metallic and superconducting materials. Kloc was a chemist. They needed to give the crystals to a physicist with a background in the physics of semiconductors.

Here the project hit an obstacle. Most of the MTSs who were interested in organic materials were already working with plastics. A lot of work had to be done to start over with the crystals, including cleaving them; choosing the right materials for the metal contacts, gate insulator, and gate electrode; and assembling the devices. There was no guarantee that the results would be any better than those that had already been achieved in plastics, and when Laudise asked Ananth Dodabalapur, one of the MTSs who was working with plastics, to work with the new crystals, Dodabalapur refused. Kloc remembered that the conversation ended with Dodabalapur making the furious declaration that there was no way he would work with the new samples.

In the freewheeling culture at Bell Labs, it would have been considered counterproductive for managers to force an MTS to do something that he or she didn't want to do, and Batlogg's thoughts turned to the possibility of bringing in someone else. A junior person would be best, someone with some experience working with semiconductors and curious about the possibility of exploring new physical phenomena. It seemed like a good opportunity for a promising young researcher to spend a couple of years in American industry, doing the measurements on Kloc's crystals that no one else wanted to do.

2

HENDRIK

Academic priorities were always important to Jan Hendrik Schön and his family. When a colleague once asked Schön about his background, he replied that he was German, and that when his mother had married an Austrian man and moved away, he had gone to live with his grandmother so that his schooling would not be interrupted. At the age of sixteen, town records show, he relocated too, joining his mother and stepfather with his brother in the Austrian Alps. They lived in Feldkirch, not far from the three-way border between Austria, Switzerland, and the principality of Liechtenstein. It was a small and insular town, wary of foreigners, and Schön soon returned to Germany to attend university.

He registered at the University of Konstanz, fifty miles north of Feldkirch on the Swiss-German border. This was a year that, as a young German man, Schön could have spent in the army, but by relocating to Austria he was able to avoid it. That was a significant step considering that Schön's father had been a soldier, but it also meant that Schön could begin university a year earlier than his fellow students. What he lacked in common background, he made up for in his manner, being much more similar to the polite and self-possessed German stereotype—he was born in August 1970 in Verden, Lower Saxony—than the gregarious Vorarlbergers of western Austria. He joined a handball team and was frequently seen running on campus in the afternoons, with his good looks, broad shoulders, and curly brown hair often complemented by elbows or knees that were scraped from a recent game.

Enrolling for physics classes, Schön was part of such a diverse intake that only about half the students in the class graduated, said a former classmate. But like other German universities, Konstanz was free and had an excellent library. "There was a good education if you wanted one," the classmate said.

Schön did. He graduated with a diploma, the German equivalent of a bachelor's and master's degree, in five years. This involved two years of foundation classes in the physical sciences and three in physics, including an education in the physics of semiconductors, the science that lay at the heart of his later fraudulent claims. The final exam was an oral, the former classmate said, and involved questions on all the physics courses. It was hard, and hard to bluff. Schön would not have had much trouble; he was intelligent, which he revealed to other students because the homework assigned in the classes was tough and was usually tackled by groups who worked together, for example, around large tables in the library, and Schön liked to help others by participating in the discussions. When a class called for a verbal presentation, he was self-assured, but he did not risk putting others' backs up by acting overconfident, nor was he what in England would be called a "swot," an intense academic type. He could move easily between work and sport, and he was usually happy to go along when invited out for a drink. Although quiet and well-mannered, Schön was not a shy person, and he could speak up or maintain eye contact when he needed to. He did not like to get into confrontations and stayed calm when others tried to engage, often preferring to think about his response and come back later rather than to react on the fly. "When something needed to be said, he said it, but he wasn't noisy," said the former classmate. Schön was interested in travel, especially in the English-speaking world (he once said that his mother had been an English teacher), and in continuing to do the science that he was good at. He had a very good memory and was often able to answer a question about physics with a moment's reflection before a relevant reference formula popped out. Taking into account what he described as a childhood that involved moves between cultures and homes, Schön's time as a student was marked by little more than a preference, and an aptitude, for fitting in.

Along with his classes, Schön's degree involved the completion of a lab-based research project. For this he applied to Ernst Bucher, a professor in the physics department who employed as many as thirty scientists working on the research and development of solar energy technology. Bucher had been in Konstanz since 1974, the middle of the world oil crisis. With solar energy as a leading alternative energy source to oil and with Germany's strong environmental movement, Schön would not have had to ask around much to realize that Ernst Bucher's lab was unlikely to be short of funding and opportunities. Along with running the lab, Bucher was a successful entrepreneur who had cofounded a local company that manufactured and sold solar panels, creating hundreds of jobs in southern Germany. He was internationally well connected and had active collaborations with research groups in South Africa, Eastern Europe, and the United States, where he had worked at the renowned Bell Labs. Bucher often arranged for his students to get internships abroad, and his lab was regarded as a gateway from Konstanz to the international scientific community.

Schön's first task was straightforward. He was assigned by Bucher to use computer code available in the lab to calculate the way in which light is absorbed by metal oxide layers for coating solar panels. A similar project was later advertised by the lab as suitable for a student "with theoretical, rather than experimental, interests."[1] The point of the diploma projects, said Schön's former classmate, was not for students to do breakthrough science, but to get a taste of research and to form a relationship with a laboratory in case of interest in progressing to earn a PhD. Schön fit in well. He joined soccer games with other students and practiced his English on the foreign visiting scientists. His results on the metal oxide layers were cautiously presented, and his published work drew heavily on the results of his computer code simulations,[2] but he was nevertheless offered a PhD project, which he accepted. Only then did some slight tension begin to develop.

Most of Ernst Bucher's students worked with silicon, the semiconductor that is commonly used not only in the computer industry, but also for making solar panels. Schön later described these students as *Silizianer,* or "Siliconites," folks working with silicon. In contrast, he was assigned to work with the complex semiconductor copper gallium selenide, which was called "CGS" (this was an alphabetical abbreviation, with "C" for copper, "G" for gallium, and "S" for selenide, rather than an abbreviation using chemical symbols). Researchers in Bucher's lab had preliminary data suggesting that CGS might be useful for ultraefficient solar cells if its properties could be modified, and this was the project assigned to Schön. He worked with crystals of CGS grown by one of the lab technicians, who fired together the elemental ingredients of copper, gallium, and selenium in a furnace. Writing after his disgrace, Schön explained to a colleague that science quickly became the only thing in his life. Although he spent time playing sports and hanging out with friends while he was in Konstanz, he also moved into a room in the nearby home of the technician who made his samples and began to orient himself increasingly to the lab. His PhD goal was to change the way in which electricity was transported through CGS. CGS is a "p-type" semiconductor, in which electric current is transported by the flow of positive charge, or holes, against a background sea of electrons. The task was to try to make it go "n-type," so that the current would be carried by the flow of negative charges, or electrons, instead. Solar panels draw their power from energy-producing junctions between p-type and n-type materials. But because the silicon that other students worked with had long been available in both n-type and p-type forms, CGS was always likely to be a side interest of Bucher's relative to the main lab effort. For Schön, this was a social as well as a scientific reality because most of the people he met at the lab worked on something different from his project. "It wasn't always easy against the might of the *Silizianer,*" he wrote in his thesis, and although he wrote with good humor, he felt that others agreed, because he noted wryly that there seemed to be a rather fast turnover of fellow "CGS-ers."

While Schön's project was independent of the lab's primary research direction on silicon, Bucher arranged things so that every student developed a core experimental skill. Much later, the rumor was spread, by Schön as well as by others, that he had spent his PhD days using a sputtering machine, a squarish oven-shaped machine for depositing metal oxide layers. Although Schön did use such equipment during his diploma work, as a PhD student he focused mostly on learning to measure photoluminescence, the light that is given out when samples of materials are stimulated by a laser beam. He took charge of a yellow-green krypton-ion laser, components for directing its beam around a table, a holder for samples of materials, and instruments for taking spectra (data that show the intensity and energy of a beam of light) from the CGS samples. Around such equipment there is usually a curtain to cut out light, so that it would have been stuffy and lonely when Schön was in the lab making his measurements. But he convinced Ernest Arushanov and Leonid Kulyuk, visiting scientists who sometimes watched or supervised him, that he was good with photoluminescence equipment, and later taught other students of Bucher's how to use it. He was also far more a careful than an intuitive experimentalist. He did not make excitable claims while he was measuring, but instead calmly collected and analyzed data, and showed it to Kulyuk, Arushanov, Bucher, and other students in the lab for help in understanding what it meant.

Given the extensive scale of Schön's later fraud, a natural question to ask is when he first started cheating. A panel of investigators convened by the University of Konstanz in 2004 stopped short of finding that Schön committed fraud as a PhD student but did conclude that he tended, even at that time, to alter or muddle his original data in order to make the presentation of his results clearer.[3] Judging by the circumstances in the Bucher lab, the stage at which this tampering could have most easily occurred was probably that between Schön's measurements, which he was seen doing and which he was competent to do, and his later circulation of data to others. This is also the stage at which colleagues who knew Schön as a student scientist reasoned, in hindsight, that he might have started to go wrong. "In my opinion at the time he was an honest and proper person. The technical work and the measurements were correct," Kulyuk told me. "But after you measure, you try to explain, to understand why the data has this shape, and that depends on your state of knowledge. So I am sure that it was after the measurements, during the analysis, that his fantasies began."

"SOMETIMES THE STUDENTS ARE AFRAID OF ANGRY QUESTIONS"

Ernst Bucher was Swiss and lived in Kreuzlingen, the Swiss town that merged into Konstanz in the south. He sometimes traveled to the lab by bus—one bus from his home to the Swiss-German border, a brisk walk over the border that was sometimes marred by inconsiderately weaving student cyclists, and a second bus up the hill from the lakefront to the lab. Like other professors at German universities, Bucher had dozens of hours of administration and classes to teach each week and relied on group leaders (research scientists who held temporary positions in his lab) to supervise the students from day to day. As a Swiss who had also worked in the United States, he also sometimes struggled with the culture. "Sometimes German students have an inferiority complex," he said. "They are afraid of being asked angry questions. You have to come and pick up their data, they don't come to you. The professors are up here and the students are down here," Bucher said, and lowered his hand down to around six inches from the floor.

Bucher told me that he made an effort to be approachable and tried to promote informality in the lab. He inspired affection and loyalty in former students I spoke with. But the extent to which he succeeded in putting everyone at his or her ease from day to day is less clear. Bucher had an expressive manner that could sound fierce and a tendency for outbursts. I heard students call him "Herr Bucher," which is not the formal "Herr Professor Doktor," but they were not so casual as to use his first name, "Ernst," as would be common for an American professor. Ironically, given Bucher's effort to avoid formalities, one student remarked to me that he came to like "Herr Bucher" once he got used to his old-school ways.

Bucher had more difficulty getting along with other faculty members in Konstanz. In 1988, the physics department moved into a new building overlooking tree canopies on the edge of the campus. Because the building was not designed to hold Bucher's industrial-scale machinery for making and testing solar cells, his lab was left behind in two inauspicious but roomy structures some way down the hill from the main university campus. When Bucher tried to create a permanent position for Christian Kloc, whom he had hired as a research scientist, other faculty opposed the move. The physics department chairman Wolfgang Dieterich explained that Kloc was highly regarded, but other faculty members were concerned that too many permanent positions in labs would reduce the number of temporary positions available for PhD students to mature into after graduation. Bucher saw this as university bureaucracy interfering with the progress of his lab. Soon afterward he lost Kloc, who was recruited by Bob Laudise and Bertram Batlogg to Bell Labs.

Such constraints may have limited Bucher's ability to ensure full supervision for students, he said. As research scientists who acted as lab mentors did not have permanent positions, they sometimes lacked authority. Bucher said that he assumed that Hendrik Schön, along with other students, received enough help and advice from older students. But there was no hands-on supervisor double-checking Schön's PhD work. In addition, several former lab members didn't see this as necessary. Bucher's lab was engaged in technology development. It was rare for there to be a sudden groundbreaking discovery that called for urgent double-checking. Instead, progress was made incrementally by several people, working as a team, through many rounds of data collection and analysis. "Do you think that a machine starts hooting when it gets a result?" one of the other students asked me. "It wasn't like that."

As in many large loose-knit labs, it would have been possible for a student in Ernst Bucher's lab to begin to stretch the truth about his findings without anyone noticing. But Bucher's lab had an important quality-control mechanism. The materials that the students worked with were often shared and could go to different experimental stations to be measured in different ways. Bucher told me that if his lab got promising results, he would have the data independently checked by sharing samples with outside labs. At one point, others remembered, there was a strong disagreement between Bucher and a group leader, Martha Lux-Steiner, who Bucher told me had shared some of the students' materials with an outside lab without permission. Lux-Steiner disputed Bucher's account, but the more important aspect is the observation that samples were intensively discussed. They were considered the property of the lab, not of individuals, and might be sent outside for independent measurements. Although some amount of low-level cheating could not be ruled out, most students would have regarded it as both risky and unnecessary to misrepresent the properties of samples in Ernst Bucher's lab.

"AUSREIßER"

Sensitive to convention, Hendrik Schön started out keeping careful track of his copper gallium selenide samples. Each piece was tagged with a label that could be used to look up how it had been made and what chemical or heat treatments it had undergone. Schön also understood that he should label data according to the sample from which he took it. He followed this practice so literally that his first published reports used labels like ER1 and ER2, which identified samples to close colleagues in the Bucher lab but would not have meant much to people outside. Schön's first data also included many fuzzy, scattered points, so that when I showed Ernst Bucher results that Schön had shown him a decade earlier, the professor read through the text, nodding, then looked at a figure, jabbed his finger at a data point that lay off the main trend, and commented, "*Ausreißer.*" The word means "outlier," a measurement that lies some way off the main trend of the data. Something jogged the equipment, or a flyaway cosmic ray passed through the sample as Schön worked, creating an unexpectedly wayward point. In German, *Ausreißer* also means "wayward child" or "maverick." Both the presence of outliers, and the scattered noisy data, suggest that Schön felt comfortable representing the flaws and idiosyncrasies in his measurement process to his professor.

But after taking data, Schön had to analyze it. Analysis involves putting the stream of numbers that come off lab instruments into equations so as to calculate physically meaningful quantities. The equations Schön used included many that were part of the well-established theory for semiconductors such as silicon and, he must have hoped, his samples of CGS. But, as commonly happens in experimental research, Schön had problems getting the results of his calculations to match the results reported by other scientists. One of his first studies started out with an analysis to compare the energies and densities of electrons in copper gallium selenide with those from other semiconductors. The analysis had three steps. First, Schön plotted points on a chart, with each point representing the energies and densities of electrons for each sample he had studied. Then he drew a sloped line, a line of best fit, to capture the overall trend of his data points. The third step was to enter the number that corresponded to the steepness of the line into a final calculation. But this result could not have been too comfortable. A paper published in 1987 by a Hungarian scientist, Bálint Pődör, summarized results from the scientific literature, all of which were higher than Schön's.[4]

The type of exercises on which Schön had excelled as an undergraduate did not involve this kind of dilemma, and in the end the number that he put into the final step of his calculation was not the slope of his best-fit line, but a slightly discrepant value, the effect of which was to bring the final result into line with the results Pődör had summarized. Relieved to have reached agreement, Schön wrote that his result "fits quite well" with those of other scientists.[5]

This tendency to fudge calculations also affected other analyses. At one point, as Schön struggled to produce a best-fit line that had the form other scientists had obtained, he felt forced to introduce an awkward ad hoc serve.[6] The result seemed to fit better with standard equations and scientific literature than it would have done without any fudging. Just as Schön liked to fit in socially and was an eager member of a sports team, so he was keen to come across as an agreeable member of the scientific community. He was not yet engaged in demonstrable scientific fraud, but he was not developing rigorous habits of analysis either, and he always tried to give a neat appearance where possible.

He also suffered from anxiety about what others might think of his work. After his disgrace, Schön circulated an email to a colleague in which he explained that he had gotten into the habit of replacing data in his publications with best-fit lines which would have given a clearer, nicer representation. He also began to smooth data out, or to connect data from different samples as if it came from the same one, fudging the sample labels that he had originally represented correctly. But he did not discuss this with anyone because, he implied, he knew that other people would not agree. "Did I tell it explicitly? No, but I shouldn't have done it to start with," he wrote. He volunteered fewer details of his analyses to others as time went by.[7] He relied heavily on other scientists' papers and, on occasion, inserted sentences or calculations word-for-word or symbol-for-symbol into his own reports, trying to make sure that he got things right.[8] And while he seemed to integrate everything he read, he rarely engaged closely with the underlying concepts. He was able to ask and answer difficult questions about physics, but he was less willing to meet a serious argumentative challenge or defend a controversial position. "Hendrik never used to argue with me. He always used to agree, say 'yes, yes, you're right,'" said a later colleague. Theoretical physicists sometimes remarked that Schön seemed to apply concepts he had learned about in a naive way, as if textbook results were familiar landmarks into which the bewildering stream of real experimental data should fit. "Schön [assumed] that if something was in a book, that thing was true," said another colleague who ended up struggling to reanalyze some of his data.

This preference for literal agreement had a devastating effect when Schön obtained puzzling experimental results. One way to make disagreements go away, he quickly realized, was to exploit experimental error. This is the idea that lab measurements usually differ from their true value owing to imprecision in the measurement process. At one point, Schön mentioned another scientist who had obtained a result different from his own, but he claimed that the other scientist's results suffered from an error, which meant that the two findings were in agreement, even though the other scientist had not reported such an error.[9] Such a commitment to paper over disagreements had the potential to annoy other scientists. "It was one thing I really didn't like about him," said one of his later colleagues. But while Schön didn't want to be disliked, it was often better than getting into an argument. Over time, his data included fewer *Ausreißer,* the outliers that did not conform to the main trend of his data, enabling him to shake off any sense of being the maverick or odd one out. By the third year of his PhD, he was beginning to cover for the difficulty he had experienced in interpreting his results by making irregular changes to his data. In photoluminescence, his main area of expertise, researchers measure the light given out by electrons moving between different energy levels in an atom. Peaks in the spectrum of light coming off a sample can be interpreted to reveal the positions of those energy levels in the material. But Schön sometimes struggled to make out any peaks, seeing nothing but a large, fuzzy background, as he admitted in his thesis. Even when one group of samples did produce some "relatively sharp" peaks, Schön wasn't able to make them out clearly enough to tell which of two possible configurations of energy levels explained them, he wrote.[10] Frustrated by the limitation, he submitted a paper to a journal soon after in which he suggested the opposite—that he could distinguish the peaks.[11] Within a month, he went further. He added a couple of clear peaks to two of his spectra[12]—see Figure 2.1. He then wrote that the numbers he derived from the data were "in good agreement with data reported in the literature."

Schön was therefore sober enough to describe an ambiguous lab experience, but revised his account later, in stages, to reach a simpler story that fit better with other scientific literature. This preference for changing measurements rather than challenging expectations

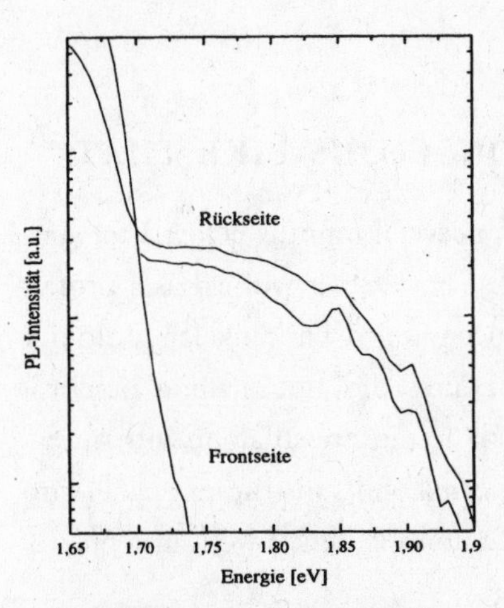

Adapted from Figure 3.30 on page 48 of J. H. Schön (1997) *Anwendungen von CuGaSe2 in der Photovoltaik.* Dissertation der Universität Konstanz, Band 329, UFO Atelier für Gestaltung und Verlag, Konstanz.

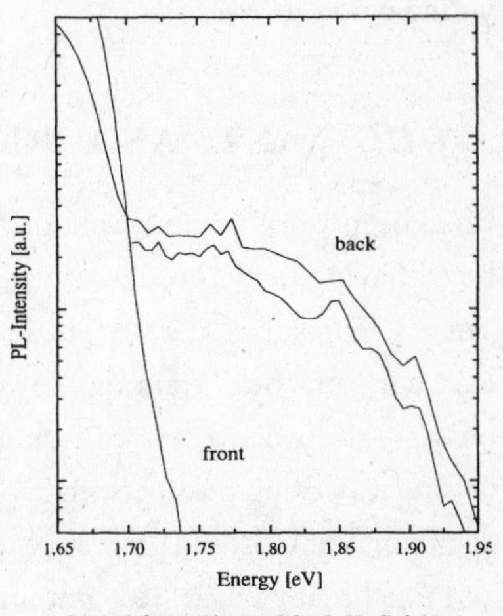

Adapted from Figure 2 in J. H. Schön, O. Schenker, L. L. Kulyuk, K. Friemelt and E. Bucher (1998) Photoluminescence characterization of polycrystalline CuGaSe2 thin films grown by rapid thermal processing. *Solar Energy Materials and Solar Cells* 51 (3-4) 371-384.

Figure 2.1: In one of his earliest manipulations, Schön made changes, adding peaks to two curves. The result was a simpler story that fit better with other scientific literature

predated any claims of astounding discoveries, supporting the sense of colleagues from the time that his later data-handling problems had little to do with excitement about things he was seeing in the lab, and more to do with a lack of comprehension and dissatisfaction with his data. There is also little evidence from people who knew Schön that he had any tendency to "see things" that others did not. He was, Kulyuk said, a "well-equilibrated young man."

In wanting his results to fit with other scientific literature, Schön was not so different from other scientists. Scientists like to reach consensus about experimental results and the theory that explains them. The difference is that Schön was so keen to avoid disagreements that his early work included almost no discussion of conflicting observations long enough that he or others might have been led to a deeper truth that could reconcile them. Instead, he forged agreements, one number or result at a time, trying to stay in compliance with science. This helps to explain why Schön did not realize—at least, not at the beginning—that he might one day get into serious trouble by faking data. If he stayed in good agreement with other scientists, how far wrong could he go?

"WHO WANTS AN UNCONSCIOUS LIKE THAT?"

I had been working on this book for several months before I remembered having made up data myself. I have never worked as a professional scientist, but as a student I once went to work in a lab. I did not have any practical training—my course did not include practical classes—and when I arrived I quickly found myself alone with an experiment, writing numbers that appeared on its instrument panel into a lab notebook. I didn't feel very comfortable, but I went in regularly and filled many pages with numbers.

Although I felt that no one else ever came into the lab, at some point the scientist who had assigned the work came in and started looking through the notebook. I remember feeling surprised that he was looking not just at roughly what the experiment had done, but at every single number. The moment I remember most clearly (this was the memory that came back and hit me eight years later) was a feeling of relief as he turned from a page of numbers that I had made up to a page of numbers that I had taken down right. Over time, I had gotten into the habit of filling out measurements according to the pattern that I thought the experiment followed. This saved time, but I don't think I was trying to get out of work so much as failing to grasp that it was important to take down every number from the instrument panel exactly as it appeared. I wouldn't have described what I did as "misconduct"—at the time, I had never heard of the concept, and felt it had not been explained to me what the standard of datataking in the lab was supposed to be, and why. That said, having remembered doing this myself, it was easier to understand that a habit of taking or analyzing data inaccurately could progress into fraud if there was never any challenge or discussion about how the data were or should be taken, and why. Habits of work rapidly become automatic, and hard to break. Think of plagiarists who, when they get caught, sometimes say that they have no memory of pilfering others' work and must have mixed their notes. One writer I know told me what happened when another writer was confronted with having plagiarized his work. He received an apology, along with the explanation that the plagiarist had acted unconsciously. He said he accepted this, but still wondered, "Who wants an unconscious like that?"[13]

Although many universities teach research ethics as part of science courses, scientists often say that it isn't until they begin work in a research lab that they understand how to apply the principles in practice. While researching for this book, I made a visit to one lab where the lab head explained how he first came to understand the importance of recordkeeping. As a PhD student, he said, he was trained to take data from his experiment into a lab notebook. One day, after working long hours on an experiment, he thought he had seen something remarkable and rushed to get his supervisor. His supervisor asked to see his lab notebook. He replied that the discovery wasn't yet in the notebook, because he hadn't had a chance to write it down. To his surprise, his supervisor was dismissive, stating that if the discovery wasn't in the notebook, it didn't count.

This helped me to understand why scientists want the numbers in lab notebooks written down with painstaking precision. Learning about or doing experiments in school, students of science study well-known phenomena and know what answer they are supposed to obtain. But professional scientists are researching into the unknown and have to reckon with the possibility that their findings will be unexpected. Making data up according to expectations would risk missing out on important discoveries, as well as on the chance to contribute new experimental facts. If, on the other hand, everything is recorded exactly as it happens, a lab notebook becomes a record of discovery in the making as well as a source of ethical conscience that can underpin future experimental claims.

Today, much of the datataking in labs is done by computers, but scientists insist that the same principles apply. Like the numbers inked firmly into a lab notebook, the data that come off lab instruments into computers is saved in a file dated with the day of acquisition, and then "never altered, and never erased," as another scientist impressed on me. The general principle is that because science is supposed to be objective, data have to be acquired and saved in such a way that human expectations, even unconscious assumptions, don't interfere. Recording the original data exactly as they happened not only makes it possible to run a reality check later; it also enables scientists to have second thoughts about anything they might have done in the course of data analysis that could have distorted the scientific message. With the documentation happening in real time, either inaccurate or dishonest datataking is more likely to be noticed, challenged, corrected, and brought to light. Even if that doesn't happen, the requirement of permanent documentation means that a lie will come back to haunt the conscience of whoever was responsible, in a worse way than a lie told in a conversation would.

At least, this is more or less the way things are supposed to work in science.

"HE DIDN'T GET IT"

When Hendrik Schön's deception unraveled in 2002, Schön told colleagues that he had never learned to save his original data. In Ernst Bucher's lab, the data acquisition was done by computer, and other students said it was clear what they were supposed to do. When they used equipment, they put in their sample, programmed the attached computer to take the data, and then, when the measurements were done, transferred the saved data files back to their own computers for analysis. Each person was responsible for saving this original data and keeping it separate from later analyses. But in one email I saw written after his disgrace, Schön insisted that, from the very beginning, he had only ever learned to save what he called "secondary data," the results of his analyses. "This is something that I have done already during my PhD thesis days. Obviously, I should have done differently, but I also did not learn it correctly (which is no excuse)," he wrote.

For saving "secondary data," Schön used a computer program called Origin that is primarily designed for analysis rather than documentation of data. Origin readily allows changes to data. It is even possible, in Origin, to manipulate data points by hand on a graph with the underlying dataset changing to match the manipulation. It is up to individual users to make sure that they use the program ethically, and keep their original data saved separately from analysis or explorations.

Schön never got into the habit of doing this, he later wrote. Even on the occasions when he happened to save his original data, he saved it only after bringing it into Origin, or turning it into a graphical image, so that in his record-keeping system, analyzed and original data were indistinguishable. Think of navigating a busy social life by keeping a dream journal and an appointment diary in the same notebook. The habitual failure to keep good records had a profound effect on Schön, who wrote after his disgrace that he sometimes misrepresented data, but added that his poor recordkeeping anyway prevented him from being able to go back and prove what had happened in the first place. "Since I do not have the original data to these experiments, I cannot prove what I have done and that I have done the experiments," he explained. So, while a skeptic once suggested to me that Schön might have deleted his original data files to cover up for not having done real experiments, Schön was able to rationalize things the other way around. Because he never got into the habit of keeping original data, and did not have witnesses to his lab work, he did not have any way to prove that he was doing them. Over time, the loss of original data turned into a source of defiance about whether keeping data would have made any difference. "How can you prove originality of data without any witness [to] the measurement?" Schön asked rhetorically, looking back after his disgrace in an email sent to a colleague.

As far as I could learn from other former members of Ernst Bucher's lab, Schön often sent data to others for discussion, but it was not common that other people asked for his data or checked that the figures he produced for published papers corresponded to it. It's impossible to know whether, if such checks had taken place, Schön would have changed his ways. But it is clear that, in a lab in which original data was not communally documented or audited, Schön did not have to ensure that the results he presented were even roughly consistent with whatever he had originally measured. The opportunity for his conscience to catch up with him was kept to a minimum, and so was the chance for anyone else to notice any changes he made. And by his own admission, when he realized he might be doing something ethically ambiguous, he didn't bring this up, and so, presumably, missed out on the opportunity to be given advice that might have helped him redirect his energy onto a more ethical path. Instead, he covered any feeling of discomfort with a neat presentation. Academic priorities were important to Schön, but showing consistency with original data wasn't part of the test that he felt he was taking.

To a large extent, he was right. As a student, Schön was judged on his detailed reports. He was a keen reader and took pride in being a fast writer too. A list of articles and talks that he compiled as an attachment to his thesis in 1997 recorded that he had published as many as twenty-four papers, with three conference proceedings and two unpublished manuscripts included for good measure. Ernst Bucher felt that much of this work was solid, as Schön had reported on trying a variety of chemical treatments in copper gallium selenide and had shown modest changes in the material's electrical properties. Yet in a lab in which samples were communally held and labeled, Schön did not go out on a limb and try to claim that he had changed copper gallium selenide from a p-type into an n-type semiconductor, the result that was the goal of his PhD. Ernst Bucher accepted this as a disappointing but authentic outcome. "He didn't get it," Bucher told me.

As a student, Schön was seen as a solid worker, but by no means as a genius. He was intelligent and had a keen sensitivity to others' results and expectations, but he was not brilliantly creative. After he became a scientific star, one lab colleague told me that at Konstanz it was speculated that his move to the renowned Bell Labs was a large part of the explanation for his success. In 1996, when Ernst Bucher was first asked whether he had a promising student who might be suitable for a postdoc position at Bell Labs, Schön's name was not the first to come up.

FITTING INTO AN EXISTING TEAM

Christian Kloc recalled that the initial idea at Bell Labs had been to find a junior researcher who could start work on organic crystals straight away. Two students in Ernst Bucher's lab had experience in doing electrical measurements of the type that Bertram Batlogg had envisioned for Kloc's crystals. But because Bell Labs was recruiting on a later and more urgent timescale than the academic year, these candidates already had other plans. Still, Batlogg remained open to considering any smart student from Ernst Bucher's lab. Those of Bucher's students who had come to Bell Labs in the past had often turned out to be as good as the students from Ivy League universities in the United States, another MTS told me, and postwar Germany was known for a strong emphasis on semiconductor physics.[14] "This was one of Germany's leading semiconductor labs," Batlogg said.

Then Schön's name entered the conversation. Schön was younger than the other students, in part because he had missed a year of military service, but also because he had not yet graduated. He was not very well known to Kloc because Bucher's lab was large and Kloc had worked more closely with other people. Although Schön's primary responsibility in Konstanz had been photoluminescence, he had done electrical measurements. Schön would also have been qualified, if not for a postdoc position, then at least for a student internship. According to one advertisement posted by Bucher's lab, the requirements for going to Bell

Labs as an intern were "understanding of semiconductors, good experimental skills, good knowledge of English, good exam results, a self-reliant and reliable worker able to fit into an existing team."[15]

Ernst Bucher told me that he had asked Schön to go. Schön seemed unsure. Bucher asked whether Schön would like go to Bell Labs for a few months as an intern. He could go back later for a postdoc position if things went well. Bucher said he thought that it might be good for Schön, who was young for his stage of life and close to his family, to spend time abroad. Schön went away to think about Bucher's suggestion, came back later, and said yes.

The internship ran from January until May 1997, when Schön returned to Konstanz to finish his PhD thesis; Bucher's lab threw a party for him. Over the next year, he went into a postdoctoral position in the Bucher lab and applied for a fellowship from the German Research Society, the Deutsche Forschungsgemeinschaft or DFG, to fund his return to Bell Labs as a postdoctoral member of technical staff. He continued working with both Bucher and Batlogg during this year. Although it would be usual for postdocs to come to Bell Labs through a formal interview process, involving a public talk, Schön was recruited prospectively, in a way that depended on how the internship went. This was a lucky break, as Schön hadn't finished his PhD and would probably have had trouble impressing a Bell Labs audience with its tame content. In the acknowledgment section of his 1997 thesis, he took care to thank Bucher not only for his PhD supervision but for the Bell Labs connection.

During the transitional year of 1998, Schön's data-handling problems began to escalate very seriously. Until 1997, Schön had compensated for deficiencies in his understanding of his data by sleights of hand in analysis, and, according to his own admission, self-reinforcing record-keeping practices. But Bell Labs was very different from low-key Konstanz. It was known for big discoveries, ambitious expectations, and a confrontational style of discussion; Schön was recruited armed with little more than a literal approach to semiconductor physics and a strong inclination to agreement.

3

A SLAVE TO PUBLICATION

Isaac Newton, the founder of modern physics, has been faulted by historians for fudging calculations on several occasions, but the case for fraud is strongest for an experiment reported in his 1704 work *Opticks.* Although Newton allegedly also fiddled the figures in his famous work on gravity, the *Principia,*[1] the theory of gravity that he put forward is still considered approximately true. In contrast, some of Newton's claims about light were influenced by a scientifically mistaken theory, and misled other scientists.

By the time Newton misrepresented his optical experiments, his motive had been festering for as long as thirty years. Newton first rose to prominence in England in 1671 after a telescope that he had built in Cambridge arrived at the Royal Society in London. Members of the society were amazed by the telescope's finely ground mirrors and sharp images, and Newton was elected to join the elite body in 1672. He followed up by publishing details of his theory on the nature of light in *Philosophical Transactions,* the Royal Society journal. The essence of the theory was that white light was made of rays and could be reflected, refracted, and split into the colors of the rainbow according to mathematical rules that Newton had formulated.

But to Newton's dismay, his report did not meet with the same unmitigated acclaim as his telescope. Both the theory and the demonstrations were attacked. One of the fiercest critics was the English physicist Robert Hooke, who was slightly senior to Newton and could be vicious in debate. Although Newton arguably defended his work successfully, he soon sank into depression. In a March 1673 letter to the secretary of the Royal Society and the editor of *Philosophical Transactions,* Henry Oldenburg, he tried to leave the Royal Society by claiming he couldn't afford the membership dues. When Oldenburg offered to waive the fees and implored Newton to stay, Newton confessed his real feelings. He was despondent about the criticism of his work and its publication had been an awful experience. Oldenburg tried to tell Newton that colleagues appreciated his work, but two months later, Newton cancelled the promised publication of lectures he had given on optics, fretting that he would be disrupted by critics if he went ahead. By 1676, Newton felt so oppressed by one self-imposed deadline to respond to a critic that he declared, "a man must either resolve to put out nothing new, or to become a slave to defend it."[2]

Newton could not bear to return to the topic of the debate with Hooke until 1704, the year after Hooke's death, and with the publication of *Opticks* he was determined to put his theory of light on a sound footing once and for all. His misrepresentation occurred while revisiting a debate that he had found distressing as a young man.

By 1704 Newton's optical theory entailed the view that it was impossible to build an achromatic lens, a lens able to produce focused images without introducing chromatic aberration, fuzzy rainbows caused by differently colored rays being refracted by different amounts. An achromatic lens would have competed with Newton's telescope, which used mirrors, rather than lenses, to correct for the undesired effect. But according to the historian Alan Shapiro, Newton's reason for thinking that an achromatic lens was impossible to build was not so much pride in his telescope as confidence in his theory.[3] Newton mistakenly assumed that when light is refracted, different colors are dispersed (seperated) in the same proportion, resulting in an even spread of colors of the rainbow. No matter how many prisms and lenses of different transparent materials were combined, separated colors could never be reunited unless a beam was bent back to its original direction, which would undo the focusing and destroy the image, Newton reasoned.

To demonstrate this point, Newton's *Opticks* contains at least one report that is very damning to him.[4] Newton claimed to have closed the shutters of his study in Cambridge, leaving only a narrow slit for a single sunbeam to pass across the darkened room to an intricate lens made of a triangular prism of glass inside a prism of water. He claimed to find that rays of white light that left the lens parallel to the original direction always remained white, while rays that were redirected were invariably split into colors.

Both this experiment and the claim that it was impossible to build an achromatic lens were accepted at the time. But in 1749, twenty-two years after Newton's death, optical scientists began to wonder. That year, the Swiss mathematician Leonhard Euler published a proposal for building an achromatic lens. In 1754, the Swedish scientist Samuel Klingenstierna showed that the match between the results and the mathematical formalism in *Opticks* could only be true if Newton's triangular prisms were set to have very small angles in them, something that Newton never mentioned. With Newton's theory and experiment under pressure, the English lensmaker and optical scientist John Dollond figured that it might be worth trying to build an achromatic lens, and in 1758 he succeeded.[5] At this point, it was concluded that Newton's experimental results had been either ambiguous or in contradiction with what he claimed. This was considered a disturbingly large error.

There was worse to come. Historians have found evidence that Newton knew that his reported results were inaccurate. In 1975 Zev Bechler pointed out that Newton's notebooks show that Newton had tried the experiment with the triangular prisms as early as 1672 and had obtained a result contrary to the one he later published. Newton also seemed to be struggling with the possibility that his real result might be interpreted to support Hooke's theory instead of his own.[6] In 2002 Alan Shapiro reported on his discovery that in an early Latin draft of *Opticks* Newton had described the questionable experiment as idealized and only in a late revision added the sentence that claimed he had actually performed it. The result was a covert and scientifically wrong misrepresentation that delayed the development of achromatic lenses by half a century, until Euler, Klingenstierna, and Dollond came to the rescue.

Isaac Newton was a proud man who wouldn't have wanted to be discredited, even after his death, so what made him declare something that he had found ambiguous with such confidence? Shapiro's work on Newton's papers has shown that, over the years, Newton became convinced of the theory supported by the misrepresented experiments and probably believed his results he reported were "the right results." But why, for the sake of integrity, could Newton not at least have stuck with the Latin draft of his book, which described the results as ideal rather than real? As it turns out, the answer to Newton's misrepresentation lies at least partly in the stiff standards that had been introduced for reporting scientific experiments, which were first articulated by the experimentalist Robert Boyle.

ROBERT BOYLE'S STANDARDS FOR SCIENTIFIC PUBLICATIONS

In their book *Leviathan and the Air Pump,* Steven Shapin and Simon Schaffer describe how the conventions governing the system of scientific publication were invented in the seventeenth century by the wealthy gentleman scientist Robert Boyle.[7] Boyle's starting point, the historians explain, was an obsession with the problem of unreliable information. In the introduction to his 1661 book *The Skeptical Chymist,* Boyle gave the example of a ship coming into port with sailors on board who tell wild and wonderful stories about faraway lands. Who knew whether the stories were true? The answer, Boyle said, was to create an elite community of people who understood the importance of separating speculation from fact, a scientific community.

With the founding in 1660 of the Royal Society, Boyle applied himself to showing others how members of a scientific community should behave. He organized demonstrations of his air pump, a gigantic contraption of glass, wood, brass, and leather that he was using to explore the properties of air. Boyle hoped that the presence of multiple witnesses would make the facts he established more certain. But because large demonstrations were not always practical, Boyle also put forward the idea that scientists should describe experiments in enough detail that others could replicate them elsewhere, increasing the number of potential witnesses to experimental facts. This principle of reproducibility is the gold standard of acceptance for scientific claims today, even though very few scientists know that it was pioneered by Boyle. The purpose of reproducibility is to put experimental facts on a firm basis.

Many ideas about scientific writing follow from the principle of reproducibility.[8] Published accounts of experiments should be candid, honest, and precise, said Boyle. The shortcomings of instruments and equipments should be described in detail. Sources of error and confusion should be explained. Failure should be carefully documented and published, the same as success. Boyle, whose own writing is very hard to read, also stressed that it didn't matter too much if scientific papers contained long, convoluted sentences or technical jargon. A scientist's top priority was precision, not style.

Demanding as this was, Boyle gained followers. Scientists all over Europe were soon building air pumps and exchanging reports about them. Reports on other natural phenomena were also circulating in Europe, often with individual scientists acting as editors and translators of the most interesting findings. By 1665 the system was formalized with the founding of the first scientific journals. The diplomat and scientist Henry Oldenburg founded *Philosophical Transactions* in England, and the writer Denis de Sallo started the *Journal des sçavans* in France. The motivation for journals was in part to provide scientists with a venue in which to stake priority claims, particularly over foreigners, but the standards set by Robert Boyle implied more than this. Publication was a duty. Those who failed to publish were excluded from membership in the scientific community, no matter how grand their technical claims. For example, one group that refused to participate were the alchemists, who were interested in turning non-precious metals into gold. Given the state of chemical knowledge in the seventeenth century, this goal was not as unreasonable as it seems today, but alchemists were very secretive, in part because in England their work had been outlawed by authorities who feared that they might disrupt the economy if they succeeded.

In describing Boyle's contributions, Shapin and Schaffer made the case that his legacy was far more than just the air pump, but also the establishment of the conventions governing modern scientific communication. Boyle also had a profound influence on the young Isaac Newton. Newton had a strong interest in alchemy, religion, and the occult, but when it came to science he knew full well that he needed to provide experimental facts to support any theory he wanted to advance. In the scenario favored by the historian Alan Shapiro, Newton added the insistence in a final draft that claimed his two questionable experiments in *Opticks* were real because he knew that without it, his book did not contain any demonstration of one part of his theory and would always be regarded as incomplete.

Boyle anticipated that scientists might feel pressure to show experimental success. To counter it, he suggested that they try to be modest and report only things that they knew for sure, rather than feeling obliged to advance whole new systems and theories—advice that Newton studiously ignored. But the system of scientific publication has more or less followed Boyle. To this day, the most respected format for communicating science is not books or treatises that pretend to solve everything, but journal articles: concise reports of specific experimental findings.

The first scientific fraudsters understood these conventions too. In 1769, a report on pendulum experiments that appeared to disprove Newton's law of gravitation was published in the French *Journal des beaux-arts et des sciences.* It was exposed as fictional in 1773. In 1784, the director of the Observatory of the Knights of Malta, Jean Auguste d'Angos, published a report of a comet in *Journal des sçavans* that was exposed as fraudulent in 1820. The Maltese were very keen to try to put themselves on the scientific map of Europe at that time, and in 1788 a second Maltese noble claimed to have discovered a new sea creature, but he published drawings that turned out to be imaginary. Two of these instances were described by Charles Babbage in his 1830 essay, "Reflections on the Decline of Science in England, and on some of its Causes."[9] Although Babbage's essay is best known for its early discussion of scientific fraud, the broader context was his sense that the scientific standards had suffered as science had become more organized and professional incentives had begun to play a role. For example, Babbage lamented the problem that some scientists seemed to hold back on criticism of others entirely because they hoped to gain in popularity and be elected to important positions in the Royal Society.

Such considerations also played into the difficulty of addressing suspected fraud. Although the standards for taking scientific data have evolved over the past two centuries, the idea that it is wrong to make things up was well established in the seventeenth century, and eighteenth- and nineteenth-century scientists were upset when instances of dishonesty came to light. At the same time, scientists have always struggled with the question of how to confront potential wrongdoing. Robert Boyle had suggested that scientists should be wary of questioning others' motives because, he feared, people might be put off reporting their experiments honestly if they felt they could be attacked for their findings. So, the first scientists were not naive. They knew that others might be unreliable. They understood that colleagues might lie, yet they cultivated an aversion to accusations because their goal was the creation of a civilized community in which they would all be free to focus on experimental facts.

"'PUBLICATION' IS THE SAME WORD AS 'PUBLICITY,' ISN'T IT?"

The Industrial Revolution brought science and technology directly into the lives of ordinary people, and in the late nineteenth century, the first popular science magazines sprang up. In 1869, *Nature* was founded in London with the primary mission of informing the public about developments at the cutting edge of science. In 1880, *Science* was founded in Washington, D.C., with the goal of helping scientists across America to communicate; it was then acquired by the American Association for the Advancement of Science, an organization that sought to promote and organize science. *Scientific American* was founded in 1845, and *National Geographic* was founded in 1888. These magazines were concerned not only with communication among scientists, but with the communication of science to the public. By the twentieth century, most scientists were no longer gentleman scientists but were hired and funded at the taxpayers' expense, often based on the number and profile of the scientific papers they had published.

Then came the Internet. Today it is possible to get a handle on the importance of a scientific paper even without understanding it—by looking online to see how many times it has been cited by other scientists. *Nature* and *Science* use password-protected web sites and email to get the word out about discovery claims on weekly timescales to science reporters, and specialist scientific journals have followed in their footsteps. The scientific literature is now a stream of current events, with research projects that have been pursued with caution and care for years turning into a news event on the day of publication. One physicist whose work was propelled into the limelight by a *Nature* press release told me: "I guess 'publication' is the same word as 'publicity,' isn't it?" Many scientists enjoy the attention, but some feel uncomfortable when they see colleagues' eyes glazing over and going "ping" with dollar signs at the public relations value. One senior European professor told me how once, when he was mentioned in the newspapers in connection with having a paper in *Science,* he received a note of congratulations from a dean at his university who he knew was not interested in his work, but who apparently saw a possibility of using the press coverage to impress people who might give the university money. "We used to publish to impress colleagues in our field. And now who are we publishing for? Deans! Students! Parents of students! Politicians!" the professor expostulated.

What senior scientists find irritating, younger scientists perceive as pressure to publish. Although "pressure to publish" is often mentioned in passing, it's worthwhile to examine briefly how it affects the kind of science that is done. Here's an assistant professor I interviewed who did not yet know whether he would get tenure—be given a permanent position: "In my department, there are some older faculty who look at the implications of your work and try to judge its long-term impact. But more and more people just look at the number of times you get cited." The effect of the trend, he said, was pressure to move into hot, fast-moving areas of research where publications would bring more attention to his group and accumulate citations, rather than the kind of science that he was most curious about and that he believed would have the most lasting value. A research scientist in his second year of a temporary position at a national laboratory in the United States told a similar story. He had applied for a large research grant. He had been told that if he got it, the lab would be able to make his position permanent. He had also begun writing up some of his data. His manager found out and suggested submitting the write-up for publication, so that they could say in the grant application that they already had a paper under submission. But the paper wasn't ready. In the end, the manager backed off, but the pressure could have resulted in a half-baked publication.

Such examples show that the quality of the scientific literature hangs in the balance between incentives and pressures to publish and scientists' sense of responsibility for the facts that they contribute. For three centuries, this sense of responsibility, and the discretion of journal editors, was regarded as sufficient to prevent too many low-quality claims from swamping the scientific literature. But in the second half of the twentieth century, journals were deluged with more and more submissions, and editors began to introduce formal quality control checks.[10] These days, most editors send submissions to outside experts for advice before deciding whether to publish them, a process called peer review. The hope is that peer review will weed out publications that aren't original or that contain poor reasoning or ambiguous results. These are either rejected, or sent back to the author with suggestions for how to improve.

But peer review has created its own set of pressures, as scientists feel that they need to present data to an ever-higher standard. One person who has thought hard about this problem is Mike Rossner, the editor of the *Journal of Cell Biology.* Rossner has received significant attention in recent years for being one of the first editors to introduce routine checks on the authenticity of images submitted to his journal; he instituted this practice after stumbling across an example of manipulated data. He has since found that as many as a quarter of scientists made some kind of after-the-fact improvement to their results, such as using image editing software to remove imperfections. In some cases—Rossner has said in around 1 percent of cases—the alteration to data is so extensive that the science involved has been misrepresented, raising the specter of scientific misconduct.

Although scientific fraud goes back to the origin of science, the concept of misconduct is fairly recent. It was not until 1999 that the United States, which is ahead of many other countries, introduced a standard definition. This says that research misconduct is the "intentional, knowing, or reckless . . . fabrication, falsification, or plagiarism" of data or text. Often, the definition of misconduct is only applied when problematic data are inserted into a manuscript or made a part of an official record such as an application for funding so that the fact or possibility of publication lies at the heart of the modern concept of misconduct. Being found guilty of research misconduct can have dire consequences for scientists, among them being sacked, sent to jail, disgraced as a cheat, or forced to publish corrections to scientific papers. Each year, the Office of Research Integrity at the U.S. Department of Health and Human Services publishes from half a dozen to a dozen official findings of misconduct, and it often bans the guilty parties from applying for government funding to do science in the future. In physics and engineering, the number of cases of misconduct found each year is unknown, as agencies that fund this research, such as the Department of Energy and the Department of Defense, do not operate offices of research integrity and rarely publicize cases when they occur.

The introduction of formal sanctions for misconduct reflects a profound change in the scientific community. As the number of scientists has grown and the publication system has become closely linked to financial and career rewards, the data that lie at the heart of the enterprise have taken on a dual role. Everyone, from journal editors to funding officials to scientists, insists that data are sacrosanct. Without data, the scientific community would be little more than a far-flung fleet of sailors with fantastical tales to tell. On the other hand, data are also currency that determines who gets published, who gets funding, who gets a permanent job, and whose conclusions are thought valid.[11] Suppose, then, that a scientist has understood that science involves data but hasn't bought into the idea that his or her data should be authentic. Over time, engagement with the system of scientific publication could push an already fraudulent scientist to ever greater levels of excess.

"MY RESULTS AREN'T PUBLISHABLE"

Of the many New Jersey towns that Hendrik Schön could have lived in, he chose Summit, a small, wealthy town a short drive from Bell Labs' Murray Hill. Summit wasn't as large a change from the prosperous university towns of southern Germany as America could have been for Schön. The town had a few walkable blocks with boutique shops and delicatessens, and Schön moved into an apartment in a yellow clapboard house with other German-speaking postdocs and students at Bell Labs. They lived on Summit's very own Park Avenue, a hilly residential street that led toward the town center in one direction and toward the highway and the airports back to Europe in the other. Schön settled in, buying a second-hand station wagon to drive to work, and going out to parties or into New York City with other students or postdocs on weekends. He missed soccer, he told friends from home, but he also developed an interest in American sports like roller hockey and basketball.

Things were harder in the lab. Sometime in 1998, Schön came in to the Bell Labs cafeteria, joined one of the large round tables, and said miserably that he had been told that "my results just aren't publishable." This story could be apocryphal, as I heard it a few times but could never trace it back to anyone who was there, but I was able to confirm that Schön had difficulty getting his first Bell Labs papers into print.

As an intern in 1997, Schön began his experiments on organic crystals working in Bertram Batlogg's lab. The room was numbered 1E–318 (it was on the third floor of Building One, in Hallway E) and was just around the corner from Batlogg's office. While the communal areas at Murray Hill had a rarefied ambience, it was in the labs that the 1940s construction of the place was most strongly felt. The crisscrossing corridors and hallways were bewildering for new recruits, and many of the labs were disorientating, with small windows and fluorescent light. 1E–318 was cluttered with equipment from many researchers that had piled up over the years. But with an air of quiet confidence, Schön convinced a colleague who offered to show him the ropes that he knew his way around in the lab.

Batlogg helped Schön get started. To begin with, Schön's reserve meant that the relationship was quite stiff. Speaking in German, Schön and Batlogg used the polite form *"Sie"* to address each other, and in his PhD thesis Schön thanked Batlogg in formal terms for his mentorship in this period. Their first experiments involved turning Christian Kloc's alpha-sexithiophene crystals into field-effect transistors. They connected the crystals into circuits by attaching gold contacts and then made a third "gate" electrode out of kapton foil, a cheap material similar to the skin of shiny helium balloons, with plastic on one side and metal on the other. They used clamps to hold the plastic side firmly against the crystals while the shiny side was made electrically live. Then Schön measured the flow of current in the circuits while he varied the voltage applied to the gate electrode. Their datataking wasn't automated to begin with. Schön apparently took measurements on pieces of paper, which he stacked in loose, unbound piles near his computer and later threw away. But even with a primitive set-up, the scientific promise seemed to be there. Batlogg and Schön saw the current across the surface of the crystals change in response to the fieldeffect, and by late 1997, Schön had two papers in preparation with Kloc and Batlogg, one of which was boldly titled "Thiophene Single Crystal Field Effect Transistors."[12]

Neither of the papers had an easy time. In January 1998 one paper was rejected by an editor at the *Journal of Applied Physics* on the basis that the Bell Labs work did not seem to cast light on the physics of any process. Although the editor tried to be polite, adding that he didn't mean to criticize the paper's technical merit, the rejection could be interpreted as scathing given that part of the motivation for the organic crystals project had been to reveal new physical effects. A second submission also went through at least one round of rejection at a journal.

Schön needed more data. Following his internship, he took his work with the crystals back to Ernst Bucher's lab, where he was seen working in front of an evaporating column, a chimney-like piece of equipment in which hot, bubbling gold evaporated upward, passing through a mask, reaching the bare surface of the crystal and forming metal contacts. By attaching wires to these contacts, Schön measured the current and voltage directly through each crystal without needing to make a field-effect transistor. He obtained mostly rising lines of points—the higher the voltage, the greater the current. In the interpretation favored by the Bell Labs group at the time, a sharp rise in current with voltage could mean that charge was flowing freely in the materials without getting caught in "traps" caused by impurities or imperfections. A measurement of "trap-free" transport would suggest that the samples were very pure and would raise hopes that in further field-effect experiments it might be possible to probe the materials' fundamental limits; that is, to measure charge flowing unobstructed at the fastest speed possible in the materials. Schön was very conscious that this was part of the motivation for the research program he had joined because, along with his experimental work, he and Batlogg also prepared a theoretical paper suggesting that other research groups' measurements on plastic films were very far from nearing the limits,[13] setting the stage for the Bell Labs organic crystals' collaboration to do better.

Schön delivered. By February 1998, he had claimed to see "trap-free" transport consistently in organic crystals. He liked to repeat results from good samples to give this impression,[14] but even this wasn't enough to produce the leaps in current that he felt were required, and by September 1998, when he returned to Bell Labs as a postdoctoral researcher and resubmitted one of the papers that had been rejected for publication the previous year, he had manipulated at least one set of points, stretching the top end of a curve to show higher currents than he had previously reported—see Figure 3.1.[15] Along with the claim to have made field-effect transistors, he was now able to include voltage and current measurements made on the crystal surface, so that two lines of evidence pointed to the possibility that the Bell Labs samples were pure enough to test the fundamental limits of plastic materials, as the managers hoped.

This time, journal editors were interested. In September 1998 Schön had a run of success getting published. As a student he had been satisfied with getting papers into specialist journals read by a small subsection of the physics community, such as *Solar Energy Materials and Solar Cells* or the *Journal of Luminescence.* But physicists at Bell Labs had more ambitious priorities and, judging by his submissions, Schön had also begun to consider it important to get into bigger journals such as *Applied Physics Letters,* published by the American Institute of Physics, and *Physical Review Letters,* published by the American Physical Society. He was also learning how to write a catchy, high-profile paper; the key, he realized, was compelling data. "Coming to Bell Labs I learned that figures should be the most important part of publications and that every figure should tell a clear story. I was reminded about that many, many times," he wrote in an email sent to a colleague after his disgrace. When Batlogg suggested that Schön add insets to some of his figures to provide additional information, Schön obliged. Schön liked to take the initiative too, and at one point he

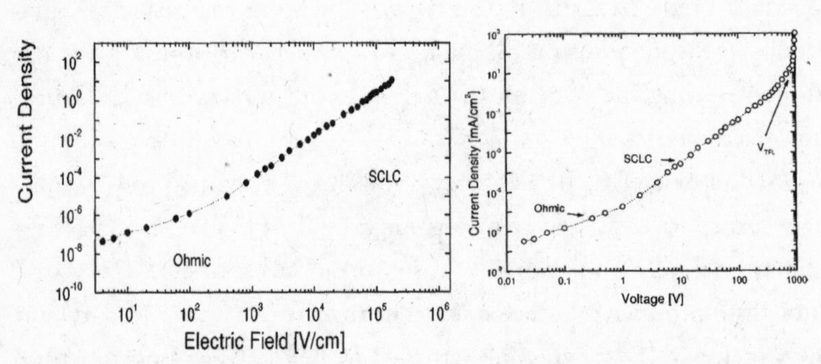

Adapted from the bottom of Figure 2 of J. H. Schön, C. Kloc, R. A. Laudise, and B. Batlogg (1998) Electrical properties of single crystals of rigid rodlike conjugated molecules. *Physical Review B* 58 (19) 12952-12957 submitted February 1998.

Adapted from Figure 1 in J. H. Schön, C. Kloc and B. Batlogg (1998) Surface and bulk mobilities of oligothiophene single crystals. *Applied Physics Letters* 74 (24) 3574-3576 submitted September 1998.

Figure 3.1: Following his recruitment to Bell Labs, Schön manipulated data, adding and moving several points to support managers' idea that samples made at Bell Labs might test the limits of plastic electronics.

used a computer graphics program to draw up a three-dimensional cartoon of the transistor made with kapton, which Batlogg happened to see over his shoulder. "I said, 'Wow, this guy's a computer whiz,'" Batlogg told me.

Schön was making a good start on pleasing his new manager, but he had more on his mind than Batlogg. In 1996, Schön had accepted the Bell Labs internship tentatively, on the understanding that he might advance into a postdoc position if things went well. But over time, he began to worry about his postdoc position developing into a permanent job. When he was asked after his disgrace whether he had had time to develop and characterize his lab methods (instead of publishing short, high-profile papers full of results and short on technical details) he wrote in reply that there were more important things and that he didn't think he would have gotten a job at Lucent for that. Other young scientists were also acutely aware that big results were the main currency. One former Bell postdoc who had hoped to get a permanent job in the exact same period as Schön told me that publications strong enough to apply for professorships at good universities were not necessarily enough to be hired at Bell Labs in this period. Bell Labs managers were looking not only for intelligence and proven expertise, but also for the possibility that a young researcher was developing an innovative line of research that had the potential to expand and to help keep the institution in the public eye. There was a growing emphasis on the journals *Nature* and *Science,* which had a broader reach than physics journals. Bill Brinkman, who managed the Physical Sciences and Engineering Division, which Schön joined, said that although *Nature* and *Science* were traditionally seen as biology journals, by the 1990s physicists began to realize that they could publish there too. *Science* had a distribution of 150,000.[16] *Physical Review Letters* had a distribution of 3,700.[17] "The game is stacked," explained Brinkman. The trend to try to get into the bigger journals affected all physicists, Brinkman thought, "but the Bell Labs manager was not immune."

As Schön began to think about getting bigger results, there was one possibility staring him in the face. During the year that Schön had spent as a postdoc in Bucher's lab in Konstanz, he had given up all hope of ever reaching his PhD goal of making n-type copper gallium selenide.[18] But by August or September 1998, around the time that he prepared to leave Konstanz to take up his postdoc position, he began to feel bolder and told Ernst Bucher that he thought he had seen n-type behavior in some of his samples of copper gallium selenide. As Schön had been working toward n-type copper gallium selenide for four years, this was a remarkable piece of apparent news. "But he didn't show me any data," Bucher said wonderingly.

Schön was floating the idea. Other students in the lab, including Markus Klenk, a graduate student working with thin films of copper gallium selenide, often discussed the interpretation of photoluminescence spectra with Schön, including some of the data in Schön's 1997 thesis. At one point—Bucher thought it was after Schön had left for Bell—Schön circulated data demonstrating the discovery of n-type conduction in copper gallium selenide. He explained, said Klenk, that the input from others had helped him with the correct interpretation of these new results, and he invited them to be coauthors on a paper reporting the result. They agreed.

Schön sent a paper about the advance to *Applied Physics Letters.* Although this work did not get a lot of attention from other scientists, Schön always felt it was an important breakthrough. As many as two years later the paper appeared at the top of a list titled "Five Most Significant Publications" that was submitted as part of a résumé for a grant application to the U.S. National Science Foundation.[19] Even after his disgrace, Schön wrote to one scientist in the field of solar energy to defend this claim in particular. Yet the project had nothing to do with organic crystals or Schön's Bell Labs collaborators. This shows that, although Schön's claims began to escalate after his first move to Bell Labs, he was not responding to immediate pressure so much as to priorities that he had learned about at Bell Labs, which he interpreted to apply across the board. His move to New Jersey might have been an opportunity to leave behind his lackluster research on copper gallium selenide and dive whole-heartedly into organic crystals. Instead, he hedged his bets, claiming an advance important to his former lab just as he was leaving. If this was an insurance policy—against the possibility of not getting hired at Bell—

it soon became more, as Schön could not have missed out on the freedom with which he could make a bigger claim when his coauthors were located elsewhere. The doubts of other scientists in the field of solar energy technology[20] suggest that this publication on n-type copper gallium selenide may have been the first time that Hendrik Schön seriously misrepresented the properties of a material. If so, it was an invaluable practice run.

"READ MY WORK!"

Whenever Christian Kloc entrusted a crystal to Hendrik Schön's care, he could expect to hear back within a few days or weeks about its mobility. The mobility is a measure of the ease with which charges are able to move through a material. Plastics typically have mobilities that are hundreds or thousands of times lower than those of silicon, which has been purified over decades of use in the computer industry. The goal of the organic crystals program was to create a form of ordered organic materials that would set new records for mobilities in organic materials. Kloc was always looking for ways to refine and improve his crystal growth recipe.

Kloc and Schön worked closely. When Kloc handed on a crystal he told Schön how he had prepared it. Schön then added metal contacts, made current and voltage measurements, and calculated the mobility. As time went by, Schön began to report higher and higher mobilities, and Kloc assumed that his ideas for improving the crystal growth conditions were working. The crystals Kloc produced looked mostly the same; they were large, colored, translucent crystals, and photographs taken by his group showed some flaws or dirt on the surface. But it was understood that the only way to detect electrical purity was by Schön's electrical measurements. Only years later did Kloc realize that some of the positive feedback he received from Schön was not right. To give one example: at one point Kloc switched from flowing argon gas to flowing hydrogen gas through his crystal growth apparatus. Schön claimed an increase in mobility, and the group formed the view that hydrogen was the best gas to use.[21] After the fraud came to light, Kloc and other researchers went back, checked the assumption, and decided that it was not clear that hydrogen was a better growth medium. But at the time, it seemed natural to attribute any improvements claimed by Schön to the latest refinement.

In August 1998, Bob Laudise, the director who had recruited Kloc and coestablished the crystals program with Batlogg, died, which left Kloc as the only crystal grower coauthoring with Schön. On the outside, their group also came into contact with Norbert Karl, a distinguished crystal grower at the University of Stuttgart, Germany, who was once described to me as "the grandfather of organic crystals." Karl had long operated very specialized crystal-growing equipment in his lab in Stuttgart and, according to his successor, Jens Pflaum, once spent as long as nine months recycling the same molecules to create a superbly pure crystal many centimeters in size. Some competitive tension emerged between Karl and the Bell group, and in 1998, when Kloc spoke at a conference on crystal growth in Germany, Karl stood up and berated him from the floor, saying "Read my work Kloc!" apparently furious that Bell Labs had failed to credit some of the older Stuttgart papers. Kloc was shell-shocked, and went back to Bell Labs with a story to tell about careful citation.

Hendrik Schön was due to speak at conferences himself over the coming year and could only have regarded the prospect of being confronted by a distinguished German professor with aversion. He went online to look up Karl's old papers[22] and started reading. The Stuttgart group had done some stunning work in the 1980s measuring the mobility of electrical charge in crystals of naphthalene. They had reported that by cooling the crystal they could produce dramatic enhancements in mobility, as would be expected if the motion of charge was governed by fundamental forces, rather than impurities and imperfections. At Bell Labs, Schön was measuring the mobility of charge on or near the crystals' surface rather than through the interior. But the citation-worthy data were of natural interest to Schön, and from this point on the form of his results and those of Norbert Karl's were in very good agreement.

As Schön's first round of measurements were beginning to come out in well-regarded physics journals, he began to show a second round to collaborators in which he echoed the findings that the Stuttgart group had reported for the interior of crystals but transferred these into the context of measurements made on the surfaces of the Bell Labs samples. The dramatic increase in mobility at low temperatures resulted in astoundingly high numbers, thousands of times greater than had ever been measured for organic crystals. Although at least some of the results showed mobilities greater than Karl's, and later turned out to be fabricated,[23] the form of the data was so similar that the numbers involved seemed reasonable to experts in the field of organic electronics. The fakery was a stroke of scientific-literary genius that put Schön's collaborators over the moon about the remarkable mobilities apparently achieved in the latest batch of Bell Labs crystals.

"CONGRATULATIONS!"

By early 1999, Schön was working on a manuscript reporting on solar cells made from organic crystals. This paper, under submission at *Nature,* didn't have an easy time, but when the editor, Karl Ziemelis, finally faxed to say that *Nature* would accept it, Batlogg added a note of congratulations at the bottom in loopy blue ink before passing the letter to his collaborators "Hendrik" and "Christian." By this point, everyone on the team was on friendly first-name terms.

Such effusive congratulations, typical as they were in the ambitious Bell Labs environment, reverberated strongly on Schön's low-key sensibilities. When managers expressed enthusiasm about the idea of submitting the next paper to the journal *Science,* along with a possible press release, Schön confided in a coauthor, Steffen Berg, his feeling that his managers' reaction was overblown. The way Berg understood this, Schön was responding strongly to the positive feedback, but was trying to sound low-key. While managers had reason to be happy not only about the paper, but also about the scientific progress they assumed had been made, Schön, who as the next chapter will describe, knew he had faked the data,[24] did not feel much empathy. He tried to play the work down, telling Berg that it had not been his first choice to send the paper to *Science*. This apparent coyness notwithstanding,

Schön had begun to realize that it might be easier to fake blockbuster results than to sculpt authentic data into submissions for specialist physics journals. Such results would also be likely to interest mainstream journals. Having had this realization, all that he needed to take science by storm were some good ideas for things to discover next.

4

GREATER EXPECTATIONS

Two years after Hendrik Schön first arrived at Bell Labs, his data handling had progressed to include some serious misrepresentation, but he was still relatively unknown in the outside scientific community. The escalation of his fraud to this next level began in 1999, as Schön settled into his position as a Bell Labs postdoctoral member of technical staff (MTS), and became involved in corporate activities at Lucent. In January 1999, for example, Bertram Batlogg, Schön's manager, sent the staff in his department a list of "stretch goals" that they might achieve through their research over the coming year. The goals had been provided to Cherry Murray, the director of the Physical Research Lab, although Batlogg made clear that they were not to be taken too seriously. Bill Brinkman, Cherry Murray's manager at the time, told me that managers understood that research was an uncertain business and did not expect the stretch goals from one year to match up to the accomplishments of the next. Still, MTSs were aware of the way their research had been pitched to senior management at Lucent Technologies, and Hendrik Schön regarded everything with the same unblinking literalism.

Of Batlogg's stretch goals for 1999, two were relevant to Schön. One was the goal of understanding organic materials as well as silicon, one of the staple materials in the electronics industry. The other was the goal of using organic crystals to create a two-dimensional electron gas. Schön knew what a two-dimensional electron gas was: it was a state that formed when electrons were free to move around on the surface of a semiconductor. Unfortunately, the transistor that Schön had made with Batlogg in 1997 was too primitive to fulfill these goals. Although semiconductors are the best-known constituent of computer chips, the insulating layer that prevents the surface of the semiconductor from touching the gate electrodes and short-circuiting is also crucially important. In the primitive device that Schön had made in 1997, the crystal was insulated from the gate electrode by a layer of fat, cheap plastic, which leaked.[1] This would have limited the performance of the devices, even if the organic crystal was capable of more.

To get to higher fields, and to accumulate enough charge for a two-dimensional electron gas to form, Schön needed a higher-quality insulator, and it wasn't long before he came across an apparently very good way of making one. One MTS, Bruce Van Dover, had recently made a high-quality insulator that it was hoped might be useful in Lucent Technologies' next generation of high-capacity memory computer chips.[2] Van Dover's insulator was made of a blend of several metal oxides and could hold seven times as much electrical charge as the standard insulators used in the computer industry. In March 1999, Bertram Batlogg designed an exhibition poster to be shown at an industry event accompanying the American Physical Society meeting. Batlogg happened to display Van Dover's data next to Hendrik Schön's latest measurements on organic crystals. After the conference, the poster was brought back to Murray Hill and put up by the lab where Schön worked, on a corner opposite the men's bathroom. Several MTSs recalled the poster, and because it was located in a place that Schön passed almost every day, he could not have missed the mesmerizing juxtaposition of his own data with the work on new insulators.

At the very least, Schön had good reason to know by March 1999 that the possibility of using an oxide layer as a performance-enhancing insulator was both something managers thought was possible, and something that could help to achieve goals for research. Over the summer, he tried to see what he could do. He scheduled a visit to Europe, vaguely mentioning that he had a conference to go to or wanted to see his family. He was still keeping up his relationship with Ernst Bucher's laboratory in Konstanz, and he sometimes mentioned that there was equipment there that, as he put it later, "no one else was using."[3] In particular, Konstanz had a sputtering machine, the same general type of equipment that Van Dover had used for depositing the blend of metal oxides to make a very effective insulator. In September, once back at Murray Hill, Schön went into the lab of Gordon Thomas, a physicist in Materials Physics Research, and spoke to Steffen Berg, an intern recruited by Batlogg who was working with Thomas on organic crystals. Schön told Berg that during the recent trip he had made a ring oscillator, an electrical circuit in which the current rises and falls in a regular clock-like way, out of field-effect transistors made from pentacene crystals. Schön showed Berg a printout of data from the new circuit.

To get this data honestly, Schön had to have made several technical advances at once. First, he had made field-effect transistors, using the Konstanz sputtering machine to sputter an insulating layer of aluminum oxide onto the surface of the pentacene crystals. Then, by applying a voltage to a gate electrode attached to this layer, he had transformed the pentacene beneath from an insulating to a conducting state. Although pentacene is usually a p-type material, in which electricity is transported by the flow of positive charge, Schön said that he had been able to change the material from p-type to n-type by varying the voltage on the gate electrode from negative to positive, influencing the organic material through the insulator. This result, called ambipolarity, had made it easier to wire the transistors into inverters—logical circuits able to reverse the direction of an incoming signal—and from there into the more complex oscillating circuit. Schön was not only studying the properties of organic materials but actually making prototype organic-crystal electronics.

Three years later, Schön revealed to his investigators how he had done it. Rather than following the method others assumed—building prototype devices, making measurements, analyzing the measurements, and refining the devices—Schön was doing science backwards. He started from the conclusion he wanted and assembled data to show it. Investigators called this methodology "falsification" because it was unclear where the data Schön showed had come from.[4] Although Schön did have several wired-up pentacene crystals in the lab in Murray Hill and in Konstanz, his own writing suggests that he was taking his inspiration rather from his understanding of other scientists' expectations. For example, Schön knew that ambipolarity, the transport of positive and negative charges, was something that colleagues at Bell Labs expected might be seen in ultrapure samples because a paper published by another MTS, Ananth Dodabalapur, had mentioned the prediction in a footnote. Schön had read Dodabalapur's proposal closely, so much so that he fixated on a reference to the history of ambipolarity written by Dodabalapur and copied it into a manuscript that he was preparing about his new circuits.[5]

Falsifying data according to these expectations saved Schön from the tricky manual task of working with and understanding the behavior of real prototype devices. But the falsification he was engaged in was hard work of a different kind. Schön spent much of his time, from early in the morning to late at night, working on one of the computers in 1E–318, barely touching his voltage supply, samples, or electrometer. Other people noticed this. "The couple of times I went into the lab, he was always sitting at his computer. I never saw him taking a measurement," said Thomas. But there was equipment relevant to these measurements and samples all around, so it seemed possible that Schön was efficient, or, in line with the hard-working all-hours culture at Murray Hill, doing his electrical measurements at night. This possibility also occurred to Schön, and coming up to the end of 1999, he decided that he would stay up all night. Batlogg had said that he was thinking of leaving Bell Labs, which meant that Schön would be out of a job if he did not manage to convert his temporary position as a postdoc into a permanent position as an MTS. On the evening of December 22, Schön did not go home but worked at Murray Hill until the morning. Close to the holidays, the research buildings were empty through the early hours, but by this point Schön did not miss the companionship.

On his bench were several organic crystals, wired up with gold contacts, sputtered with aluminum oxide, and decorated with a gate electrode. One of these went into the cryostat, a flask for keeping a sample at low temperature while making electrical measurements. The experiment Schön had in mind involved pumping in liquid helium from a separate container, a dewar, waiting for everything to cool down, then increasing the voltage on the gate electrode and applying a magnetic field to the sample, while making measurements of the electrical resistance of the crystal's surface. Tuned by the gate, the amount of charge on the surface was supposed to increase, causing the resistance to fall. If a two-dimensional electron gas formed, then it would be possible for quantum effects to emerge, and the resistance might begin to fall in sudden quantum drops, rather than continuously. Although Schön did not understand the details of quantum theory,[6] he did know that research groups who had formed two-dimensional electron gases in materials like silicon and gallium arsenide had measured a rippling staircase of quantum drops. The Nobel Prize–winning team that had first measured the quantum Hall effect in 1980 in Germany had seen this staircase in resistance on a silicon crystal at 1.5 degrees above absolute zero.[7] Schön reckoned his tetracene crystal at 1.7 degrees, a couple of tenths of a degree warmer. His all-night run resulted in minutely detailed data showing the discovery of the quantum Hall effect in an organic crystal. This was a marvelous advance that agreed with previous findings and that demonstrated the potential of organic materials to follow in the footsteps of traditional semiconductors. From Schön's perspective, he had reached a second of the department's 1999 "stretch goals" just in time before Christmas.

On the morning of December 23, Schön was exhausted, bleary-eyed, and unshaven when Christian Kloc came into the lab to see how things were going. Kloc saw cold condensing water on the cryostat and dewar from the low temperatures reached during the overnight measurement run. He did not doubt that Schön had been up all night measuring. Then Schön showed the data. Kloc was thrilled to see the incredible quantum effects appearing in the devices and said that they should get Bertram Batlogg and show him. Schön claimed later, in an email to a colleague written after his disgrace, that he had called Batlogg at home early that morning and asked him to come into the lab right away, but such an excitable call would have been out of character for him, and in fact this was not something Kloc remembered Schön mentioning at the time. Instead, Schön simply agreed that it would be a good idea to search for Batlogg. Down the main corridor of Building One they went, looking in labs and offices for Batlogg, including in the office of Jeannie Moskowitz, the department secretary. There were other MTSs in the secretary's office, sorting things out before the end of the year. Schön showed them his results. Blown away by the brilliantly clear, noiseless rippling curves, the other MTSs said the result was "fantastic." Schön also showed the data to Horst Störmer, which was a very good test because Störmer was the co-winner of the 1998 Nobel Prize for his work on the quantum Hall effect in gallium arsenide. At the Christmas party later that day, Störmer also told others the result was "fantastic."

But with the holidays and the search for Batlogg as distractions, nobody went back into the lab and looked at the experiment. Schön later mentioned that he had taken the crystal out of the cryostat very promptly because he was returning to Europe and wanted to take it with him, which he could have figured other MTSs would understand. After all, Bell Labs had a precedent for dramatic year-end breakthroughs. On December 23, 1947, the Nobel Prize–winning team of William Shockley, John Bardeen, and Walter Brattain had demonstrated the first transistor on a germanium crystal at Murray Hill. The 1997 book *Crystal Fire,* a history of the transistor part-funded by a corporate grant arranged by Brinkman, and often recommended by Batlogg, told the story. In the book, Shockley was quoted calling his experiment a "magnificent Christmas present" for his research group and for Bell Labs.[8]

Schön might have been all alone in the lab, but in another sense he had very good company. And he liked to fit with convention. Any one incidence of agreement—between the department stretch goals and his results, between the history of Bell Labs and his timing, or even between a footnote in a paper he had lifted a sentence from and his data—might not have meant much on its own, but taken together it was part of the same pattern of fitting in. After some time in the United States, Schön got into the habit of sending "Happy Holidays" emails and later told a reporter for *Science* that his discovery of the quantum Hall effect had been "a nice Christmas present."[9]

EXCEPTIONALLY HIGH QUALITY SAMPLES

Bertram Batlogg later described the escalation of the case as imperceptibly gradual. He said that after the fraud was exposed, he replayed everything in his mind, trying to find a time when he had been presented with two clear alternatives, and chosen one that had allowed Schön to continue, when he should have chosen another. It did not happen that way, he said. That said, Schön's relationship with Batlogg has to haven taken a fateful turn some time in the second half of 1999 when Schön first came back from Konstanz claiming to have laid down a gate insulator made of aluminum oxide. From this point on, no one at Bell Labs understood how Schön was getting his results.

The general idea was the following. Schön put each crystal into a sputtering machine, a squarish machine with a gun inside that was able to fire a beam of ionized particles. He put in a target made of aluminum oxide. He closed the lid and set the machine to pump out air. Then he activated the gun, causing ionized particles to hit the target so that particles of aluminum oxide sputtered everywhere. When he took the crystal out, its surface was covered with a thin layer of aluminum oxide. Finally, he added a gate electrode and applied a voltage to coax the surface of the organic material into the desired electrical state.

In the year from late 1999 until late 2000, Batlogg communicated elements of this picture to at least three outside scientists I interviewed: George Sawatzky and Teun Klapwijk, who spoke privately with Batlogg, and Leo Kouwenhoven, who saw Batlogg speak in public. All three gained the impression that there was a sputtering facility for depositing aluminum oxide at Bell Labs, Murray Hill. Sawatzky said that Batlogg recommended he buy brand new, ultraclean equipment if he wanted to do similar work. Yet this picture was at odds with what Schön claimed to be doing. First, Schön claimed to be working on sputtering in Konstanz, not at Murray Hill. Second, Schön was using the sputtering machine in Ernst Bucher's lab that he had first used as a diploma student in 1993 and that he had used to sputter a coating of indium—tin oxide—onto organic solar cells for his first Bell Labs *Nature* paper with Kloc and Batlogg in 1998.[10] Connecting the dots would have led colleagues to the conclusion that Schön was treating the ultrapure crystal samples by putting them into a sputtering machine that had been used for other materials. The sputtering had to be a dirty process, not an ultraclean one, and it was not taking place at Murray Hill.

But the dots were not connected, almost certainly because of assumptions that dated back to managers' motivation for setting up a research program on organic crystals in the first place. Batlogg and Laudise had started out with the idea that by growing high-quality crystals at Bell Labs they might improve on results that others had obtained in plastics. Schön's data appeared to suggest that the program was working. When asked how organic materials, which are traditionally rather dirty, could support such impressive results, Batlogg told another department head, Dick Slusher, that Kloc had been making very pure samples. He told an MTS working with silicon, Don Monroe, that the organic crystals were structured in layers that could be cleaved to form clean surfaces for transistor experiments. Christian Kloc did not dispute this, as he had been working for at least a year with feedback from Schön suggesting that he really was making ultrapure crystals. As for Schön, he tended to agree with whatever others assumed about his work, even when one piece of information contradicted another. The collaboration wrote in their first *Science* paper that the deposition of the insulator did not introduce any impurities or imperfections to the crystals.[11]

Over the coming months, Batlogg's grasp of the details of Schön's laboratory technique could only have become more tenuous, as Batlogg made arrangements to leave Bell Labs for unrelated reasons. In late 1999, Batlogg was offered a job as a professor at the Swiss Federal Institute of Technology, which he accepted. With some humor, he bought a new tweed jacket, befitting an academic. He also learned that he had won the American Physical Society's $5,000 David Adler Lectureship Award for Materials Physics, largely on the basis of some of his earlier work. To mark the occasion, he was asked to address a large plenary session at the March 2000 meeting of the American Physical Society in Minneapolis, where he undertook to present some of Schön's breakthrough results for the first time. The talk was an illuminating tour through many of the important breakthroughs of twentieth-century physics—including the quantum Hall effect—transported into the twenty-first-century setting of the colorful organic crystals. "Whatever you would have predicted, that was what the next result was," said David Muller, an MTS in the Department of Materials Physics Research who came across Batlogg preparing his slides in the printer room at Murray Hill before the meeting.

Batlogg's exposition at the American Physical Society meeting opened the floodgates to invitations. Colleagues wanted him to come and show the exciting new data at conferences, university departments, and summer schools all over the world. His schedule became very hectic. Batlogg could be a formidable presence at conferences, not only when he spoke but when he sat in the second or third row, taking longhand notes on a large notepad and asking other speakers difficult questions. He was now on occasion seen sitting in others' talks while hurriedly preparing his own forthcoming presentation on his laptop.

Not being present on the ground, Batlogg relied heavily on the sense that experiments were going on at Murray Hill. At the same time, Schön's data began to gain external credibility from Batlogg's support, plunging the collaboration and the scientific community into a dangerous feedback loop. When Batlogg gave a seminar at the Kavli Institute at the University of California, Santa Barbara, in August 2000, he opened by saying how wonderful it was to come out to the coastal location in the bright Pacific sunshine because in New Jersey it seemed to rain two days out of three. He gave generous credit to Schön and Kloc, his collaborators back at Murray Hill, for the work that he would present. But he also invoked his own past record on high-temperature superconductivity, explaining, "it's the same me," and adding that he had gone beyond his "first love" to work with organic materials. This association had a devastating effect, as outside scientists began to weigh the data from Schön, who was relatively unknown, with the full force of Batlogg's reputation. At one point, when pressed by a skeptical questioner, Batlogg slipped into the first person to explain the workings of part of an experiment supposedly completed by Schön. "I inject the charges here,"[12] he said, apparently referring to the experimental setup. Batlogg took a didactic tone toward his audience, explaining the lab technique in such general terms that others assumed he had to be simplifying relative to what he knew. When the results seemed to go down well, Batlogg was reaffirmed in his positive view of the situation.

In principle, data acquired by other MTSs might have added some nuance to the picture derived from Schön's data. Although

Schön tended to claim that his field-effect samples were in Konstanz, pre-empting the possibility of independent measurements, several scientists at Murray Hill were able to obtain and measure unprocessed crystals directly from Christian Kloc. Andrea Markelz, a postdoc, acquired terahertz ray spectra from Kloc's alpha hexathiophene crystals and said she got "really pretty" textbook data. Theo Siegrist, a materials scientist, found by scattering X-rays through Kloc's pentacene that the arrangement of molecules was different from that previously reported by other researchers. Both results supported the idea that Kloc was making novel high-quality crystals and were not in contradiction with Schön's spectacular claims. But the work of other researchers was not an independent check because the group did not pin down the relationship between electrical measurements made by Schön and the measurements made by the others. "We were on the periphery," said Gordon Thomas, who sometimes worked with the crystals too. There was also little effort to bring everyone together into a closer team, partly because, with the total number of scientists working in research at Bell Labs in decline, many of the MTSs working with the crystals left Murray Hill for other reasons around the time that Schön's claims were beginning to escalate. In addition, data from other researchers tended to be less of a priority than Schön's. When reviewers at one journal asked for changes to the paper reporting on the structure of pentacene crystals, the Bell revision took so long that the editors reset the clock, and another research group published some of the same results in the meantime.[13] A second paper on alpha-sexithiophene was sent to the journal *Physical Review B* but later withdrawn by Thomas because he felt that the results didn't add enough to existing science on organic crystals to justify publication.[14] The only person to measure the electrical properties of the crystals apart from Schön was Mark Lee, an MTS with a background in the study of conventional semiconductors like silicon. Lee was mostly interested in studying the vibrations in the crystals, but when he made electrical measurements similar to Schön's by a different method, he got less impressive results, a fact that was mostly addressed in a footnote in the paper published about his work, on which Batlogg, Kloc, and Schön were also authors.[15] Clearly, data that were ambiguous or ordinary were delayed or sidelined, even as Schön's remarkable results were expedited into print. This wasn't a conspiracy, said Siegrist. Rather "Hendrik's data were more exciting," simpler, easier to interpret and to write up. Schön was fast at generating manuscripts, and Batlogg had learned, during the formative rush of interest in high-temperature superconductivity, to be aware that when he began to feel excited about a new result, unknown competitors might be ahead and about to get into print first.

Batlogg had always been a skilled presenter, but he had been known for keeping a close eye on the science too, maintaining enough restraint to ensure that his group's claims were reliable. Horst Störmer, who followed the collaboration, advised a colleague by email in 2001 that while Batlogg and Schön were working together on the interpretation, Schön was doing all the experiments. Störmer's assumption was that Batlogg was too busy to spend time in the lab. Many senior scientists did the same, delegating the task of experimentation to junior colleagues.

But the situation was more complex than the stereotypical situation of a senior scientist who is too busy to spend time in the lab. At a university, a junior researcher such as Schön would work in his supervisor's lab until at some point he graduated and moved away to establish his own research effort. But at Bell Labs the supervisory relationships were looser. After it became clear that Batlogg was leaving Murray Hill to build up a new lab in Zurich, Schön began to inherit space in 1E–318. Schön also took the additional precaution of carrying out a key experimental step in Konstanz, which put additional distance between himself and his collaborators. And while he seemed very self-sufficient in the lab, Schön kept colleagues busy with requests for other kinds of help. He often asked for help understanding scientific concepts, and for checking references, the choice of which other scientific literature he ought to rely on. Once he had grasped his colleagues' understanding of the framework of an experimental question, he could easily produce data that appeared to them to answer it. Batlogg liked to feel in command of experimental concepts and always enjoyed receiving Schön's latest graphs. Batlogg spent a lot of time talking with and guiding Schön, but he told me that he had noticed his younger collaborator did not respond well to angry questions or interference.

Schön was sensitive and always produced his best results when treated with gentle, hands-off encouragement.

Schön's results became even more remarkable as he began to receive helpful suggestions from outside Bell Labs. In December 1999, Schön had a fifteen-minute presentation slot in a session at the Materials Research Society fall meeting, which is held every year in the Back Bay district of Boston, Massachusetts. One of the people in his audience was Robert Haddon, a chemist at the University of California, Riverside, who had previously been at Bell Labs. After the talk, Haddon went up to Schön and talked to him about experiments pioneered at Murray Hill in the 1990s. Haddon's group had worked with carbon-60, soccer ball–shaped molecules of sixty carbon atoms, also known as buckyballs. Carbon-60 conducts electricity, but Haddon's group had gone further and turned it into a superconductor by adding potassium to supply extra electrical charge.[16] Since then, Haddon's group had tried using the field effect to draw large amounts of charge onto the surface of a buckyball sample and induce superconductivity without the need for additional chemicals, but they did not have any success.

Schön was interested in Haddon's story and about two weeks later sent word that Christian Kloc was starting to grow large crystals of carbon-60. In the New Year, Schön got in touch again, to say that he had made a field-effect transistor from the crystals. He floated some numbers, saying that he had seen superconductivity at a temperature of eleven degrees above absolute zero. Finally, he sent the data, smooth curves of resistance that audaciously fell to zero as the temperature of the crystals was lowered. "I was shocked," said Haddon, thinking about the fact that his group had tried a similar experiment without any success. "I thought I was a very bad experimentalist. If he could do that in a month or two how bad must I be? I had the idea, the carbon-60, and the transistors and I didn't get it? I was very disappointed. But I said 'Wow.'"

Haddon figured Schön must be technically competent and have good control of his experimental setup. Schön's resistance curves did not look at all like the kind of data that might have arisen from a sloppy experiment. They had a larger-than-life, doubt-dispelling quality. Schön knew this because he later told investigators that he had used an equation to calculate a very smooth sweep of data in order to avoid doubt.[17] When scientists doubt others' claims, they tend to ask detailed questions about the method that has produced them—questions that Schön must have known he would have struggled to answer. But the smooth data helped to stall inquiries. When Don Hamann, another department head, commented to Batlogg that it was surprising that the group could obtain the high mobility of charge needed to produce the quantum Hall results in an organic material that, unlike silicon, had not been refined in purity over many years, Batlogg pointed Hamann back to the compelling data, and there seemed little more to say. Schön also won a bye, at least initially, on a crucial question. Although he claimed to be using aluminum oxide as his gate insulator, superconductivity required that the electric field on the gate electrode be ramped up so high that his layer of aluminum oxide would be expected to rupture.

To begin with, Schön wasn't even aware of this possibility.[18] His close colleagues did not focus so much on his lab technique as on the intrinsic properties of the crystals that he claimed to be able to tune with that technique. After the fulfillment of Haddon's suggestion went down well, Schön proceeded to report superconductivity in crystals of pentacene and tetracene, which had never previously exhibited the effect. Like many falsifiers of data, Schön paid attention to falsifying the physically meaningful signals in his data, but often neglected the noise, the details that were peculiar to individual experiments.[19] So did the close colleagues whose expectations he was working to fulfill. Batlogg, for example, did not notice that the results of different experiments were sometimes duplicated, and he was even proud of an apparent clash between the consistency of the results that Schön was getting and a classic reference book on organic electronics by the chemist Martin Pope and the physicist Charles E. Swenberg.[20] According to the reference book, different organic materials should exhibit different electronic behavior. Batlogg suggested (both in a recommendation letter he wrote about Schön's research and while speaking in Santa Barbara) that this was only because previous experimenters had been working with dirty materials that contained a wide variety of impurities. In light of the breakthrough progress at Bell Labs, the book by Pope and Swenberg would need to be updated.

And new ideas kept coming. At one point an MTS in the lab next door, Harold Hwang, told Schön that it was problematic to claim superconductivity only on the basis of a drop in the resistance of his crystals, because of the possibility that an experimental artifact such as a short circuit might produce the same effect. Schön did not argue, but simply asked Hwang what would be needed convince him. Hwang described a couple of follow-up experiments, and Schön duly came back and claimed to have performed one, leading to another publication in *Science.*[21] Peter Littlewood, a theoretical physicist at Cambridge University who had previously been the head of the Department of Theoretical Physics Research and spent summers at Murray Hill, described the situation this way: "You would say something, and then it would happen. You would get caught up in the progress of the subject. For a long time, I didn't believe it could have been fraud because I didn't believe one person could make all that up. Then I realized, we all made it up." Unknowingly.

By July 2000, Schön had reported such a variety of phenomena in organic materials that he, Kloc, and Batlogg shared the $10,000 Industrial Award at the International Conference on the Science and Technology of Synthetic Metals in Bad Gastein in Batlogg's native Austria. Batlogg gave a plenary address about the team's work at the meeting. After he spoke, Fred Wudl, a fellow Austrian, an ex-Bell Labs MTS and an expert in organic materials at the University of California, Santa Barbara, stood up to declare amazement. The Bell Labs group had been able to tune the electrical behavior of organic crystals radically, changing them from insulators into conductors and even into superconductors. This was, as one researcher described it, "like turning an apple into an orange," a feat as remarkable as that of the alchemists who three hundred years earlier had tried to turn base metals into gold. Normally, Wudl pointed out, physicists who wanted to study such a wide range of effects would have to approach chemists and ask for help creating materials with the desired electronic properties. But because the properties of solid matter arise from the interaction of electrons with other electrons, atoms, and molecules, it was, in principle, possible that an electric field might force a material to behave as if it were chemically different. Wudl said he told Batlogg, "You're going to put chemists out of a job," and Batlogg didn't deny it.

If Schön's various experiments hadn't worked as predicted, Batlogg said that he or others at Bell Labs might have tried sooner to focus on the aluminum oxide gating technique Schön claimed to be using. If something someone had suggested had failed, it might have been natural to ask what, exactly, in the details of the method caused some experiments to work, and not others that were predicted to follow from them. But everything worked. "If something's working, you tend not to ask why it's working," Batlogg said. Such a streak of success might sound too good to be true, but Bell Labs' managers were configured to think differently. Thomas Palstra, a former postdoc of Batlogg's who had gone on to be a professor of solid state chemistry at the University of Groningen in the Netherlands, explained: "Bell Labs management thought long and hard about research strategy. They had a keen eye on new developments. On young people. It was not left up to chance." As Schön delivered results in line with expectations, managers were led, not to suspicion, but to the conclusion that the research program on organic crystals was well conceived, and that Batlogg had recruited the right people to make the conception into a reality.

GOOD LUCK FITTING WITH MY DATA

Benzene, the most well known of the aromatic molecules, was famously first described in 1865 by the German chemist August Kekulé. A benzene molecule is a ring of six carbon atoms, held together by a glue of orbiting electrons. The organic crystals grown at Bell Labs were made of one or more similar rings with electronic orbits that overlapped, producing elaborate pathways, along which electric current could flow. The pathways were far more complex than the wide bands available to electrons flowing through the monolithic interior of a silicone crystal. In organic crystals, the moving charge was expected to drag against the molecules' loopy electronic orbits. Technically, the carriers of charge were not considered to be individual electrons but large clouds of distortion, called polarons.

Hendrik Schön knew about polarons. They were described a textbook he was reading about organic crystals.[22] The textbook warned that there were several theories that could be used to describe charge transport in organic materials. But Schön did not register the pitfall. In late 1999, he fit his data on the transport of charge through organic materials closely to a theory of polarons in the textbook that had been conceived by Theodore Holstein, a mid-twentieth-century physicist. But when it came to the blockbuster breakthroughs, Schön fit his data more closely to expectations arising from discoveries made in conventional materials such as silicon. In following two theories, one novel, one more conventional, Schön began to produce contradictions in the heart of his data.

Theoretical physicists were the first to notice. In the culture at Bell Labs, it was expected that whenever an experimentalist had new data, he or she would pass it around to see what others thought. It wasn't long before Batlogg showed Schön's data to Peter Littlewood. Littlewood, along with others, quickly realized that some of the numbers associated with Schön's spectacular breakthroughs were different from what might be expected. To take a simple example, in December 1999 Schön had reported measurements of the quantum Hall effect at a temperature of 1.7 degrees above absolute zero. Later, Schön said that some of the data also revealed the fractional quantum Hall effect, in which the fluid fractured into exotic particles. The more exotic result is usually only seen at a few thousandths of a degree above absolute zero, thousands of times colder than the temperature that Schön had announced for the experiment. To Littlewood, the high temperature underlined how surprising it was that Schön had been able to measure such high charge mobilities in organic materials.

Keen to understand more, Littlewood began to examine Schön's earlier data showing how charge was transported through Kloc's organic crystals. The earlier data fit with Holstein's theory, which Littlewood knew was favored for situations in which the polarons were very light and nimble. This seemed to make sense because Schön was claiming he could also see dramatic effects like superconductivity and the quantum Hall effect, which also required mobile charges.

But other calculations suggested that polarons in organic crystals should be much heavier than those needed for the Holstein theory to apply.

The apparent contradiction led to a number of lively arguments between Littlewood and Batlogg. Christian Kloc, being a chemist rather than a physicist, was not present in many of these discussions. Schön was, but he did not say much. Throughout this period, Schön struggled to understand the science of polarons, and when he spoke at the American Physical Society in Minneapolis in March 2000 and someone asked what physical mechanism enabled the charge to be so mobile, Schön stared at the questioner blankly, until David Dunlap, the session chairman from the University of Minnesota, came to the rescue with a possible answer, Dunlap recalled. As Littlewood and Batlogg argued at Murray Hill, Schön sat calmly, listened, and took notes. Littlewood later said that Hendrik Schön was the best listener of any experimental physicist in the world.

To try to get to the bottom of the problem, Littlewood asked Schön for his original data. It was not a good time for Schön to be asked. He had submitted several high-profile fabricated papers to journals and lacked original data to back some of these up. Whether for this reason or another, he began to become visibly worried about recordkeeping for the first time. He told Kloc and apparently also his friend, intern Bernd Woelfing, that he had run out of space on his computer. He spent a day or two transferring data and files to another computer. (Two years later, when investigators asked for original data, Schön was going to say that he had deleted files after running out of space. This suggestion was believed by colleagues who remembered his computer crisis, but was given short shrift by others and was also inconsistent with Schön's insistences, in email to a colleague sent after his disgrace, that his failure to keep original data dated back to his PhD days.)

Fortunately for Schön, Littlewood was happy when he received a small batch of "original data" showing measurements of current moving through crystals at different electric fields and temperatures. Schön sounded friendly and wished Littlewood good luck fitting his theory to the data. The measurements he sent contained plenty of scattered points and noise and looked realistic. The mobility of the charge was unusually high (Schön may have taken some data, but later tampered with it, as investigators later found that he had fabricated at least some of the curves derived from similar measurements[23]), but Littlewood set to work, writing computer code to try to analyze the numbers.

Littlewood also assigned his postdoc at Cambridge, Misha Turlakov, to try to see whether there was any way to apply known theory to explain the apparent contradictions. But before that could be explored, Turlakov inadvertently provided Schön with a way out of the predicament. On a visit to Bell Labs, Turlakov asked whether it would be possible to get a close handle on the polarons in organic crystals by measuring the electrical properties of the crystals at intermediate temperatures. The only measurements Schön had reported that were directly sensitive to the mass of the polarons were at low temperature or high temperature, and Turlakov figured that an experiment that worked in the middle range would be a quick way to double-check whether the apparent inconsistency was real or the result of a mistaken analysis as Littlewood thought.

As Littlewood grappled with the data, Schön's counteroffensive was under way: a new batch of "analyzed" data along the precise lines Turlakov had outlined, which pointed to only one possible conclusion. The contradiction with theory was real. There were polarons, and they were behaving as Schön's early data suggested they might. They were uncommonly light and nimble at low temperatures and heavy at high temperatures. Initially, Littlewood queried the results, but around three weeks later received more data from Schön that supported the conclusion, and backed off. "This completely destroyed my objections . . . and forced me to look for a new theory," Littlewood said.

Because of his Bell Labs connection, Littlewood had made an unusual effort to meet the team halfway in understanding the new data. But he was far from the only theoretical physicist to question their claims. In December 2000, when Batlogg gave a talk at a superconductivity meeting in British Columbia organized by the Canadian Institute for Advanced Research, he was approached by a trio of theoretical physicists: Boris Spivak, Nikolai Prokofiev, and Philip Stamp. Each had a question for Batlogg. Like Littlewood earlier in the year, the three theorists were struck by the ways in which Schön's data fit well to theoretical equations—but to different equations from those that they expected to apply. And more people were concerned about these issues than Batlogg realized. At least one Schön manuscript about charge transport was rejected for publication by a reviewer who used an argument of Stamp's. Although, by convention, the reviewer was anonymous, this did not prevent Batlogg from approaching Stamp at a later conference to express displeasure, apparently having assumed that Stamp was the referee, Stamp said. Stamp was not the reviewer and did not know what Batlogg was talking about, but he later spoke to the Canadian physicist George Sawatzky, who told Stamp that he sympathized with his theoretical arguments and had used one in a recent review.[24]

Batlogg's response to the theoretical objections was that, given a clash between theory and experiment, he had to stand by his collaboration's data. If the data were not consistent with theory, then new theory would have to be developed, he argued. Stamp's view was that the amount of new theory needed was too much for the data to be plausible. The discussion was not resolved at the time, but by 2001 Schön was quietly making progress in working up results that extended his earlier theoretical anomalies into apparently strong evidence for a direct clash between experiment and theory. Batlogg was delighted. After months of impressive but plausible findings, it seemed that the team had finally stumbled across a phenomenon that had never been seen before. The possibility that polarons could change their nature so radically was more than a breakthrough. It was a truly fundamental discovery.

By following expectations as he naively understood them, Schön had landed himself and his collaborators in a bind. Either he stood by and allowed theoretical physicists to unravel the story, or he pushed on, hoping that by generating enough data he could throw the theorists off the trail. Schön chose the second option. It was not an easy path. Along with the rejected manuscript, another of Schön's papers on charge transport had trouble getting into print.[25] But the polaron manuscript may well have been headed for a positive reception eventually. It contained several graphs of compelling data, was strident in claiming the discovery of a new physical phenomenon, and was submitted to *Science.* It might have appeared in print, to tremendous excitement and controversy, some time in 2002. That did not happen—not because a problem was uncovered with the polaron data, but because, in the meantime, Schön began to branch out in a different, and even riskier, direction.

5

NOT READY TO BE A PRODUCT

Lucent Technologies made money by selling electronics and communications equipment to telephone and Internet providers like Verizon and AT&T. Although the dot-com boom inflated Lucent's share price, it also made for a very competitive market. Lucent fell behind, including in its marketing of the all-important routers that are used to switch signals on the Internet. Its chief executive officer, Richard McGinn, leveraged Lucent's large size to buy up smaller companies, in the hope of swallowing up products in development. That put pressure on Bell Labs, which, as Lucent's research and development organization, was supposed to be coming up with products internally. Bell Labs engineers were pulling out all the stops to develop new technology for the sales teams to market.

Traditionally, Bell Labs scientists contributed to product development in one of several ways. Some worked on products. Some advised engineers who worked on products. But sometimes, the science done by physicists had no obvious relevance to product development. These staff and their managers were always under pressure to argue that their work might be technologically relevant in the longer term.

One way to do this was to file patent applications on research. Lucent paid a salary bonus of $1,000 to any member of technical staff who filed a patent application, irrespective of whether the application was successful. Some MTSs made thousands of dollars this way each year, and even those who did not find the money much of a motivator still regarded patent filings as useful evidence that they were innovating for the company. By 1999, Bell Labs was applying for more than six patents a day and earning as many as four, a fact that company executives advertised to investors in the press and in filings to the U.S. Securities and Exchange Commission.

This could have been a good investment for Lucent if all the pending patents had involved research on useful technologies. But before that could be tested, something happened that profoundly changed the circumstances for science at Bell Labs. The dot-com bubble burst. Up until the end of 1999, Lucent had insisted it was financially healthy, but in January 2000 the company suddenly admitted that it would make less money than projected in the first quarter of the year. Its shares dropped in value by nearly 30 percent in a day,[1] and Bell Labs managers woke up to the realization that their honeymoon with Lucent was over.

After buying up smaller companies, Lucent now seemed too large. In March 2000, the company announced a spin-off of its corporate networking business into a new company, called Avaya. Rumors began to circulate that company executives were planning to split Bell Labs down the middle. Those plans became official in July 2000; Lucent was going to sell off its microelectronics business as a new company, called Agere Systems, and focus on communications technology. Physicists and engineers who worked with silicon electronics, or electronics with other traditional semiconductors like gallium arsenide, would be spun off and lose their Bell Labs affiliation. Lucent planned to keep others, including physicists working on astrophysics, micromechanical machines or MEMs for routing optical signals, and organic electronics, Hendrik Schön's field of expertise. By 2001, Lucent had also sold its fiber optics business to Furakawa Electric of Japan and Corning in the United States.

These sales and spin-offs, beginning with Agere in 2000, led to a dilemma for managers in the Physical Research Lab. For decades, managers had argued that science was relevant to telecommunications because of the interplay between science and technology. In practice, that interplay involved discussions between Bell Labs physicists and engineers. But all of a sudden the engineers who were most likely to use the expertise of the physicists were being spun off, as were most of the information age technologies that the physicists had supported. Don Hamann, the head of Theoretical Materials Research, recalled that the question of how the Physical Research Lab would be useful to Lucent in its new, slimmed-down form kept coming up at management meetings. "One of the few things anyone could come up with was public relations," he said. By keeping up its practice of press-releasing exciting scientific findings, the lab could continue to demonstrate to investors, customers, and anyone else that Lucent had a sound, long-term technological future. Again, this was a question of survival; with revenues falling, managers had to make the argument that their jobs and the jobs of their staff were worth keeping. Around this time, Bell Labs introduced more important titles for the middle managers. "Department heads" were renamed "department directors," and the "director" of the Physical Research Lab was renamed "vice president."

The start of the Internet bust also coincided with Bell Labs' first publicity about the research effort that Hendrik Schön had joined. In March 2000, Lucent put out a release titled, "Physicists Create Organic Electronics and Hi-Speed Circuits."[2] This followed an application for a patent[3] on Schön's work, the apparent strategy in which was to use data that Hendrik Schön had reported from his electronic circuits, made from organic crystals, to stake a broader priority claim on the idea of one day making similar circuits out of plastic films, which were more industrially relevant.

Despite this appearance, Lucent managers later insisted that they had never considered Schön's work a commercial prospect. Further down the press release, the company noted that organics were only at the research stage, while the "hi-speed" circuits of the title turned out to refer to the work of an entirely different research group. "We weren't thinking of using Schön's results for usable devices. His results were too far on the research side to think of them as practical in the rapid time frame needed for developing a product," wrote Bill Brinkman, the former vice president of the Physical

Sciences and Engineering Division, in an essay published in *Physics Today* after Schön's disgrace.[4] But through press releases and patents, Schön's work contributed to the impression that the Physical Research Lab was providing longer-term technological value to Lucent, an idea that managers were very keen to promote. This was not necessarily cynical, as many of the scientific managers truly believed that science could deliver value for the company in the long term. Writing in 2001, before Schön's disgrace, Brinkman himself gave the work on organic crystals as a perfect example of longer-term research that might be technologically important one day, even if it wasn't clear how.[5] If anything, the fact that Schön was not working on a product meant that there was somewhat less pressure for him or his managers to confirm every aspect of his claims immediately. Lucent didn't have customers who had ordered a batch of organic crystal field-effect transistors and were going to be asking for their money back if they could not measure superconductivity or the fractional quantum Hall effect.

The relative unimportance of Schön's work as a technological reality was also reflected in the slapdash way in which Lucent patented his research. His first claims to have induced superconductivity in crystals of carbon-60 were filed at the U.S. Patent Office on the same day that his research paper appeared in print, which, if the patent had been a serious commercial prospect, would have risked competitors' getting wind of the idea—for example through the journal review process—and scooping the company to the intellectual property.[6] Lucent also failed to inform Robert Haddon, the University of California professor who had arguably conceived the idea, and was not invited to join the patent application. And the company irritated patent examiners at the U.S. Patent Office, who wrote to complain that the application seemed to overlap with another of Lucent's patents on plastic electronics.[7] All these things were easily avoidable, supporting the claims by managers after Schön's disgrace that his work had not been taken seriously as a potential product. But in uncertain economic times, the appearance of innovation could shore up investor confidence and was valuable in itself. In response to the Patent Office's objection, Lucent asked for more time, as Schön proceeded to run further superconductivity experiments. His data were then shipped back out to the Patent Office in an attempt to shore up an application[8] and add one more number to the total on "the patent clock," a digital scoreboard in the lobby entrance at Murray Hill that counted Bell Labs patents as if managers believed that every single one was a home run.

RESIDENT VISITOR

While managers believed that Schön's work was some way from commercial relevance, Schön was not so sure. Even as he signed off on the organic circuitry patent, Schön began to claim that he had used films made of many grains of crystals that were a step closer to plastic electronics than single organic crystals to make transistors too.[9] Yet personally, he wasn't interested in the commercial side of things. He did not see himself as a corporate high-flyer, and he had modest financial tastes. He rarely took a taxi when public transport was available and claimed few expenses despite taking many overseas trips. He preferred to dress down, wearing a plaid shirt or pullover even when invited to speak in public. When he understood that he had to dress up, he did so earnestly, donning a conventional dark suit, white shirt, and orange or red tie. At parties, he drank beer, not cocktails, and preferred to sit out rather than dance. And he was not out for a quick buck; when a friend had to leave the country and was unsure what he owed, Schön took care of a signed blank check, and filled it out correctly for the right amount.

But, as Schön wrote to a colleague after his disgrace, one thing he did consider was his permanent job at Lucent. Even though he was getting lots of papers published, the structural changes at the company had left his job situation in limbo. Bertram Batlogg's last day as director of the Department of Materials Physics Research was in May 2000. Although Batlogg continued to be listed as the director of Solid State Physics Research until November, his department had been taken over by Peter Gammel, a director who was allocated to Agere, and then by Eric Isaacs, an MTS who was promoted to be the new director of Materials Physics Research in July. As this happened, Schön's position changed from "postdoctoral Member of Technical Staff" to "Resident Visitor." Schön did not feel that the new designation came over well, and in one résumé I later saw, he simply claimed to have been a postdoctoral researcher throughout these months, even though he no longer had a supervisor and even though Batlogg had begun to refer to him as his *former* postdoc.[10]

Over the summer of insecurity, Schön focused on submitting manuscripts more than ever before. He sent out at least four breakthrough papers in June, four in July, and as many as five in August. His new manager, Isaacs, pushed hard for submissions to *Nature* and *Science* that could boost the department's profile, as far as one of the other MTSs in Materials Physics Research, Rafi Kleiman, recalled: "Eric gave a talk around that time. He had one slide that was like *Science* this, *Nature* that, *Science* this, *Nature* that." I asked Kleiman if this put Schön under pressure. Kleiman said no. Schön was producing too far over the level that managers would even have expected. "The pressure was on us, because we weren't publishing [as many] *Science* and *Nature* papers."

By responding so promptly to incentives, Schön could pre-empt any need for pressure. When he felt pressure, it was self-imposed and related to the need to avoid detection. For example, as the profile of the organic crystals work rose inside Bell Labs, Schön faced the risk that other MTSs might try to join the research area and, perhaps, form independent assessments of his claims. He responded to this threat in the way that he had learned worked best—with large quantities of fabricated data. At one point, Art Ramirez, another MTS in the Physical Research Lab, showed Schön a crystal of perylene, an organic material that Ramirez had received from a chemistry colleague. Ramirez told Schön that he was keen to try perylene as the channel of a field-effect transistor and branch out from the alpha-sexithiophene, pentacene, and tetracene that Schön had tried. Schön acted friendly but also mentioned that he had already done some work with perylene. Ramirez said that he would start the next week. But on Monday morning, Schön came in to work with a manuscript that reported on the use of perlyene to make a field-effect transistor and gave it to Ramirez.[11] To show that he was getting the same results in perylene as for other materials, he falsified the data,[12] using the same transistor curves as he had used in another paper on alpha-sexithiophene but recaptioning them for the new manuscript. Ramirez did not notice the falsification, and backed off. Ramirez, Schön, and a group of others played basketball together regularly on Friday afternoons in the grounds at Murray Hill, and Ramirez knew Schön as "a good team player" who was usually happy to pass the ball to others. This time, clearly, it was Schön's turn to score.

In getting away with this, Schön was relying on a competitive environment in which an appearance of aggressive posturing would be readily forgiven. But it was also true that oversight of manuscripts was less strict than it had been historically at Bell Laboratories. Bell Labs had built up a reputation for excellent science in part because it operated with what former MTSs and managers insisted was a thorough system of internal review. Before submitting papers for publication, MTSs wrote up their research in technical memoranda that circulated inside the company or were posted on an internal database, where other MTSs could read them. In order to be submitted for publication, the memoranda had to be signed off by managers, who traditionally wanted to see reviews from people outside the authors' department. More important than this, the availability of the technical memoranda meant that there was a lot of informal discussion around interesting results so that any problems would be noticed. Managers admitted publicly after the Schön case that the system for manuscript release fell into disuse through the 1990s. As Bell Labs downsized, it contained fewer departments to provide independent assessments of new research, and the burden of oversight fell increasingly on the department directors, who had traditionally played the role of advocating for their staff's research. Thus, while Schön was still being exposed to institutional goal-setting exercises that were typical for industrial research, he had a level of freedom to ship out manuscripts that was more typical for a university. After Batlogg left, Schön felt that his new director, Eric Isaacs, was relatively uninvolved with his research, as he did not acknowledge him for comments on many papers. (Isaacs declined to give an interview regarding his six-month appointment in managing Schön.)

Managers also had more to worry about than keeping track of Schön. The spin-off of Agere meant that MTSs with far longer records at the company were in limbo, talking anxiously with their managers about whether they would be allocated to the spin-off or stay at Bell Labs. All hiring was on hold with good people stuck on temporary contracts, and MTSs were anxious for each other. But through all the uncertainty, Schön's commitment to Bell Labs barely wavered. In April 2000 he was invited to give a talk at the Max Planck Institute in Stuttgart, Germany, by Klaus von Klitzing, the Nobel Prize–winning physicist who had discovered the quantum Hall effect in silicon in 1980. Klitzing had been shown some of Schön's data by Batlogg at the American Physical Society meeting in March, and he was enthusiastic about the quantum Hall effect in the new class of materials. Schön traveled to Stuttgart, and afterwards received a job offer from Klitzing, who wanted to set up an effort to make field-effect transistors from organic crystals in Stuttgart, where Norbert Karl, the organic crystals pioneer, was based. The offer was for a junior position, but in Germany the Max Planck Institutes are the most prestigious and coveted places for scientists to work. Taking the job would have enabled Schön to return to Germany, where he was more comfortable socially, would have been closer to family, and could have escaped the uncertainty at Bell Labs. Schön thought for a couple of weeks. Then he replied to Klitzing to refuse the offer. He said that he planned to stay at Bell Labs for another two years because his research there was going well.

THE "PIT STOP" LASER

In late 1999 and early 2000, three managers in Schön's reporting line were promoted. Arun Netravali became president of Bell Lab. Bill Brinkman became vice president of research, and Cherry Murray replaced Brinkman as vice president of the physical sciences and engineering division. This left a gap at the top of the Physical Research Lab that was filled by Federico Capasso, who had been the director of the department of semiconductor physics research. Capasso was arguably the most distinguished scientist yet to take on responsibility for managing Hendrik Schön.

Like Brinkman and Murray, Capasso was a member of the prestigious U.S. National Academy of Sciences. He had also been named a Bell Labs Fellow, an important internal honor. He was well known for his work as part of a Bell Labs team that had developed the quantum cascade laser, an ultra-high-efficiency infrared laser that was successfully commercialized. Although Capasso had never coauthored a scientific paper with Schön, he was a fierce advocate of the idea that science plays an important role in technology development, and in this regard Schön's research was one of the most promising prospects in his new lab. Capasso was remembered by one colleague I interviewed as having described Schön in glowing terms, as a young researcher with extraordinary capabilities. But while Capasso was scientifically highly regarded, his credentials to be a senior manager at a time when many MTSs were anxious about the future are not so clear. Capasso was, and still is, described by scientists who disagree with him as hard to communicate with, in part because of the zest with which he communicates his own point of view. I first saw Capasso in 2008, six years after the Schön case, as I sat in a conference session where he was due to give a public talk. He arrived on the scene talking into a cell phone, his voice carrying over many rows of his waiting audience, as he instructed a colleague that, in preparation to show some new results, they should pick out the best laser spectra and make sure to use some spectacular colors. His half-moon glasses were secured firmly around his neck with a safety cord, and he carried an air of impenetrable ebullience.

With the promotion of Capasso, lasers landed firmly on Hendrik Schön's agenda too. Much of the groundwork had already been done. Ananth Dodabalapur, one of the MTSs working with plastic electronics, had been studying light emission in organic materials since the 1990s. From Schön's earlier report of ambipolarity, in which a single organic crystal supported the transport of both positive and negative charges, it was logical to ask whether the two types of charge could be made to meet up, canceling out and channeling their energy into the production of a laser beam. In the spring of 2000 Schön produced laser spectra from just such a device supposedly built in Konstanz, and turned up in Dodabalapur's office with printouts of the data, asking for help understanding what they meant. Dick Slusher, who was director of Optical Physics Research and Dodabalapur's manager, remembered that getting the laser results had seemed to take Schön quite a while, as if he were making a pit stop from his racetrack of papers about the transport of charge in organic crystal transistors.

Schön said that he had built the prototype laser by sputtering aluminum oxide onto opposing sides of a single crystal of tetracene. He had added gate electrodes, one on each side, to induce negative and positive charges using the field effect. The two charge types supposedly met in the middle, where a coherent beam of yellowy-green laser light supposedly came out. Earlier attempts at organic lasers had required at least some light to be put into the material as a trigger, but Schön figured that electricity alone could power his laser, making his result a major breakthrough. When Dodabalapur asked to see the data, Schön sent him several files in his preferred file format, Origin. These included noisy, realistic spectra that might well have come from a laser of some kind, although Schön was later found to have fabricated at least some of them and again duplicated data from one context to another.[13] Schön was familiar with the way laser spectra ought to look, because he had spent much of his PhD making measurements of photoluminescence with Ernst Bucher's yellow-green krypton-ion laser in Konstanz.

But while Schön knew how to use a laser, he did not know how to build one, and his ignorance soon began to show. A crystal doesn't turn into a laser simply because it is stimulated to emit light. It has to be carefully machined into a cavity so that the light will be trapped inside, build up, and then exit in an intense, coherent laser beam. Most lasers include a mirrored cavity where the light is trapped, but the three-dimensional graphic that Schön circulated to colleagues didn't show this, and experts at Murray Hill were puzzled. It was assumed that the inside faces of the translucent tetracene crystals must be acting as a kind of internal reflective surface or mirror, but the mathematics of this wasn't easy to reconcile with the details of Schön's supposedly measured spectra. Questions about the nature of Schön's light spectra led to at least one meeting in Federico Capasso's office.

Batlogg remembered sitting to the side. He wasn't a laser expert, and Schön was showing data that Batlogg couldn't check easily against his own expectations. Dick Slusher was there, and so was Federico Capasso, both sitting at a small table covered with laser spectra. Ananth Dodabalapur had written out several pages of calculations, trying to reconcile the data with those that would be expected for an organic laser. This was photocopied and handed out. There was a conversation about where, exactly, in Schön's samples the light was coming out. The group decided that one further experiment would be helpful to confirm whether the light was really laser light, and they relayed the news to Schön, who, Slusher remembered was mostly off-site in Konstanz while his results were discussed at Murray Hill. Batlogg said that the ongoing discussions were closed to everyone's satisfaction, especially when Schön provided an additional plot that answered some of the questions that had come up. "We brought in additional experts and the data passed their scrutiny," said Batlogg.

Slusher remembered things slightly differently. Slusher was an older, adversarial, and somewhat grizzled department head; it was he who commented to me that he had never liked the way that Hendrik Schön always seemed to agree with whatever was said to him. The previous year, Slusher had questioned the apparent lack of noise in Schön's transistor measurements and now felt troubled by the lack of clarity about how the laser worked. He said he made his reservations clear to everyone and also took aside Dodabalapur, the MTS reporting to him who was preparing to cosign a paper about the laser with Schön, and warned him what might happen if the submission went ahead. "I said 'if you publish this you're going to get objections in the scientific community,'" he said, "'people will say 'it's not lasing.' It could tarnish the reputation of Bell Labs." Among Slusher's concerns was the fact that the laser spectra Schön said he had acquired at two different temperatures had a strikingly similar form. Usually, the output of an operating laser would be expected to vary as its temperature changed. The spectra were in fact not only similar, but exact duplicates,[14] but Slusher did not notice that. Instead, from the similarity, Slusher reasoned that Schön might be detecting reproducible sparks of light, but could not have built a laser. Other data, showing the precise relationship between the light output of the supposed laser and the current input, seemed if anything more perplexing.

But, having given his opinion, Slusher did not go out on a limb to try to prevent submission of the manuscript. Of everyone involved in the discussions, Federico Capasso was the most distinguished laser expert, the most senior manager, and the best qualified to judge. Capasso spent time inspecting the spectra, and his feeling was that whatever Schön had seen, it was something very interesting, if a bit strange, according to Slusher. Slusher remembered their conclusion being that, as part of the discussion arising from the publication of such an interesting result, scientific progress would be made. The eventual agreement was that a paper containing the results could go ahead, on the understanding that Schön would bring his samples back from Konstanz to Murray Hill as soon as possible, to enable them to be carefully measured and characterized in follow-up work. Schön, Dodabalapur, Kloc, and Batlogg wrote a paper that claimed boldly in the title that they had made a laser but noted in the text in passing that not everything was understood, and promised to publish a more detailed study in future.[15] The paper was submitted to *Science,* where it was accepted to be published in just over three weeks, which was very fast compared with the journal's average acceptance time of more than three months.[16]

As far as I could learn, the submission of Schön's laser manuscript marked the first time that several Bell Labs managers independent of Bertram Batlogg had reviewed an outgoing Schön manuscript rigorously. Despite the fact that many of Schön's manuscripts were submitted without much internal review, the laser was recognized as significant enough to warrant careful scrutiny. The result of the review was ambiguous; the data raised some serious questions. But managers rationalized the decision to go ahead with the thought that those questions could be clarified in follow-up research. Although this was a relatively new idea to apply at Bell Labs, which had a reputation for rigor, the general idea that follow-up research might be needed to clarify a preliminary result is always available to scientists working on the frontier. One former MTS even remembered Federico Capasso responding to some of the uncertainty associated with one of Schön's later papers by pointing out an example from the history of organic electronics.

In October 2000, Schön's field of plastic electronics gained a major boost with the award of the Nobel Prize for Chemistry to Alan Heeger, a physicist at the University of California, Santa Barbara, and the chemists Alan MacDiarmid of the University of Pennsylvania and Hideki Shirakawa of the University of Tsukuba, who had discovered the first electrically conducting plastics. Their contributions are usually dated to 1977, but Heeger's interest in the field had an interesting prehistory that was well known at Bell Labs. In 1973, Heeger and others had published data revealing a spike of high conductivity in rare samples of a plastic compound called TTF-TCNQ. They suggested that this might be due to fluctuations of superconductivity in plastics.[17] Bell Labs scientists trying to replicate the work soon showed that the signal that Heeger had thought was a sign of superconductivity could be explained by distortions in current flow caused by cracks forming where the metal contacts attached to the surface of some samples; the exciting new result was due, in part, to an experimental artifact.[18] Heeger's paper nevertheless prompted a huge amount of scientific interest. By 1976, additional measurements on TTF-TCNQ had come in from more than eighteen laboratories, which made it clear that Bell Labs was right. Although some of the spike was due to a real effect, much of it could be explained by the artifact.[19] Heeger dropped the claim of superconductivity, but continued to be interested in similar materials and later teamed up with MacDiarmid and Shirakawa to work with another chemical, polyacetylene, which led to the Nobel Prize–winning discovery of conductivity in plastics. In 2001, Federico Capasso was remembered giving this as an example showing that even if an early result turns out to have been misinterpreted, it can be set straight by other scientists and help to open up a new field of research. That could apply to Hendrik Schön's promising new results, too.

When Capasso applied this thinking to Schön's work, he would have been doing so on the understanding that Schön's data were real. It was the interpretation of Schön's data, not their authenticity, that was the subject of internal discussion. Capasso declined to give an interview about his thinking in approving the Schön papers for publication, but he made clear in passing conversation that he would be uncomfortable with the suggestion[20] that fraudulent work might stimulate new science.[21] The managers' review had focused on the question of whether the spectra that Schön showed were really laser light, or some other kind of light. But it was assumed that there was a device emitting light of some kind, even though, in fact, no one at Bell Labs had seen it. Schön had always made clear his laser was in Konstanz, pre-empting any possible request for an impromptu demonstration at Murray Hill.

In July 2000, when the laser paper came out in *Science,* the managers did not hesitate to take maximum advantage of the public relations opportunity. They sent out a press release claiming the discovery of a "Bell Labs laser," without making any mention of the Konstanz location or crediting Schön as its sole operator. Capasso was quoted explaining that "these research results open up a whole new set of possibilities for electrically-driven lasers."[22] The press release did not mention the existence of the internal doubts about whether the light was really laser light, nor did Slusher remember being consulted for it, but the existence of some uncertainty was covered by the clever use of the phrase "research results," which conveyed that the work was only at the research stage. And the press release did contain one clear caveat. Modern fiber optics relised on the transmission of invisible infrared light. Because the light coming out of the "Bell Labs laser" was yellow-green, the device was not likely to be useful in a communications product any time soon.

"I'M NOT TOO INTERESTED IN LASERS ANY MORE"

Even before his laser paper went out to *Science*, Schön began to hedge. He knew that others wanted him to bring the laser back from

Konstanz to Murray Hill, and he warned Bertram Batlogg that the experiment sometimes worked and sometimes didn't. This caused Batlogg to suggest to David Muller, an MTS with an expertise in using an electron microscope to image materials, that it might be interesting to get hold of Schön's samples and see if there were, for example, any cracks or damage on the surface of the crystals that could potentially explain the laser variability. Muller was not involved in the discussions that preceded the submission of the laser paper, but he asked Schön for samples some time between March and May of 2000, he said. This request was hard for Schön to avoid, because Muller's lab was next to 1E–318, where Schön worked. Schön agreed that he would bring the samples back from Konstanz, but when he next returned, he told Muller that he had forgotten. That left Muller to assume, as had Art Ramirez, that Schön was feeling territorial and did not want to get involved in a new collaboration in case he lost control of his turf.

But outside was another story. As managers had expected, the publication of the laser claim provoked the interest of many experts in the field of organic materials, and by November Schön was back in Boston at the Materials Research Society Fall Meeting, where he had been invited to open a session on organic materials with a half-hour presentation, a good advance on the mid-session fifteen-minute slot that he had been allocated the previous year. By this time, Schön had quietly replaced the duplicated spectrum queried by Dick Slusher with a new one.[23] But he got some tricky questions, particularly from Stephen Forrest of Princeton University, who questioned whether the light Schön had seen was really laser light. Word of Forrest's questioning spread among other Bell Labs MTSs at the meeting and soon reached David Muller, who could no longer understand Schön's failure to share samples because the images that Muller could have taken with an electron microscope would have helped to answer the questions that had come up. "This was something that could have helped him," Muller explained. Schön began to give Muller an increasingly wide berth, and whenever he came back from Konstanz from a trip and saw Muller, he would smile and show his empty hands before he could be asked for samples.

Samples were not a new issue for Schön, but he was becoming more cautious about showing them to others as time went by. In 1999, Schön had let Steffen Berg see organic crystals that had been fried or destroyed by the high electric fields he had applied in the course of his experiments. He had showed Berg fragments of crystals, samples with large, blackened burned-out holes where he had attached gold contacts and increased the voltage. But when Ernst Bucher, Schön's former PhD supervisor, asked for samples of Schön's n-type copper gallium selenide, Schön said no. Bucher reasoned that his former student was having so much success that he had become arrogant. One person who had some chance of tracking samples was Christian Kloc, who knew at least how many crystals he had given Schön to measure in the first place. Schön made an effort to try to keep Kloc on board, asking him at one point whether he could claim to have seen superconductivity when he had only measured the result on one crystal. Kloc said he did not know, but he introduced the question as a discussion topic in the Murray Hill cafeteria and came back with the conclusion that they could do this, provided that their manuscript noted that only one sample had been studied. Judging from his papers, Schön preferred not to take that option and so had to keep the rate of his breakthrough claims below the rate at which materials were provided to him by Kloc. By the time Schön gave his job talk in 2000, he was so consumed by the need to pre-empt requests for samples that when one MTS asked what happened when the direction of the voltage in one of his made-up superconducting devices was switched from positive to negative, Schön answered carefully that unfortunately the sample had burned up and self-destructed when he tried that.

The other MTSs accepted this explanation, but the laser was harder to defend than the reports of superconductivity. It was not something that Schön had been able to handle through a staged overnight run of the type he had performed for the fractional quantum Hall effect. The invention of a new type of laser at Murray Hill would almost certainly have prompted demonstration requests from other MTS. But Schön had always made clear that the experiment was off-site and did the same with his next claim, the world's first light-emitting transistor. The idea came from Ananth Dodabalapur, who had proposed the idea in a paper in 1996[24] and suggested that Schön might use the ambipolar crystals to try it out. The idea was that Schön could use the field effect to induce positive and negative charges on different regions of the surface of an alpha-sexithiophene crystal. The two charge types would then flow toward each other, giving out light where they met. Such a device might be useful in optical communications or even as a light source, similar to the cost-effective light-emitting diodes or LEDs that have made their way into all kinds of products from bicycle lamps to traffic lights. So unlike the laser, which had started out with Schön producing data that he then shared with others, leading to lengthy discussions about how to interpret it, the light-emitting transistor had been conceived by Dodabalapur. Schön's data came later and fit closely enough to Dodabalapur's expectations that there was less discussion about whether the interpretation of the data was correct; when Dick Slusher reviewed the paper,[25] he found it plausible. Having fit his claims better to colleagues' expectations, Schön did not have to reach any new understanding about sharing samples.

But that did not make the questions about the laser go away, and Schön ended up having to change his plans for further publications. He had originally intended to write more about the laser and told one reporter that he hoped to tune it to emit light at different wavelengths,[26] something that would have made headway in lifting the caveat that the Bell Labs press release had placed on the possible future commercial development of his research. But until he brought samples back to Murray Hill, he could not make further progress in claiming new results without bringing up the old discussions. Even worse, Dodabalapur began to ask if he could come to the lab in Konstanz and see the samples there, adding that when the laser was next operational, Schön should call Dodabalapur, and that wherever Dodabalapur was in the world, he would come right away to see it. Schön later told a colleague that Dodabalapur had been the only person interested in seeing his experiments. But, Schön added, the laser experiments were difficult, and by that time he was not too interested in lasers any more.

He was not too interested in having others work on the laser either. Schön's laser paper was vague about how his samples had been prepared, something that made it hard for others to reproduce or build on the work. After the paper was published, Marc Baldo, Stephen Forrest's student at Princeton, calculated the amount of power needed for tetracene to start lasing and came out with a number 1,000 times higher than the one Schön reported. Baldo had trouble opening a proper discussion with Schön about what would explain this. Then, when Schön received a copy of a draft of a paper that discussed the laser from Forrest, Schön asked if he could share it internally at Bell Labs. Forrest said yes. Some time later, Schön got back to Forrest with nothing more than the conciliatory remark that maybe he should have tried harder to make tetracene lase. But Bell Labs did not make this climb-down publicly known, and Marc Baldo's work faced a critical reception before being accepted for print in the journal *Physical Review B.*[27]

Schön (almost certainly) felt very uncomfortable with his difficulty sharing samples and involving others in his experiments. By nature, Schön liked to be helpful. He offered to sort out a computer when he saw his former supervisor, Ernst Bucher, having trouble with it. He let a colleague's parents into Bell Labs through a side door on the weekend, holding open the outside door with an entertaining bow. He helped a student intern with school coursework. He liked team sports. He would have loved to share samples with other scientists, if only he could, but, as Ernst Bucher put it, commenting in hindsight, "I think he couldn't." If only Bell Labs managers had set a higher bar for the submission of the laser paper (for example by requiring that the samples be brought onto the premises at Murray Hill so that the follow-up work that everyone agreed was needed could begin) Schön would not have been able to deliver, and he knew it. Bell Labs might have avoided press-releasing a major fraud, even if it wouldn't have uncovered one.

But events were taking a different course. Schön's data evidencing the emission of light provided an opportunity for managers to put out a press release that illustrated the vibrancy of the Physical Research Lab, at a time when there were questions about the future of science at Bell Labs. Despite Schön's evasiveness, it was assumed that others would follow up on his work with work of their own, and that scientific progress would result. In this way, by mid-2000 the thinking was in place to allow for the possibility that the correct interpretation of Schön's results might not come to light at Bell Labs' Murray Hill, but somewhere on the outside.

6

JOURNALS WITH "SPECIAL STATUS"

The journal *Nature* is headquartered in central London, in a converted warehouse on an alleyway behind the train stations at King's Cross St Pancras. Daylight floods into the building through a huge glass ceiling and mezzanines to the desks of the manuscript editors several floors below. The most coveted view is that from the newsroom, the windows looking down onto a stretch of green canal where painted narrowboats dock and a semicircular tunnel entrance leads under the roadways.

Hendrik Schön's first frauds arrived by mail. Together with hundreds of other incoming manuscripts, they were placed in "the new pile," and decorated with a cover page to warn editors that they had not yet been evaluated. *Nature* was, and is, a commercial enterprise, owned by the privately held company Macmillan Publishers[1] and with a worldwide circulation of about 60,000.[2] Its mission, published in 1869, was "to place before the general public the grand results of scientific work and scientific discovery" and to give scientists "early information of all advances made in any branch of natural knowledge throughout the world."[3] Working with something like this in mind, *Nature*'s editors began the evaluation process for each incoming manuscript on as fast a turnaround as they could. By the end of every business day, or perhaps by the end of Tuesday if there had been a backup over the weekend, "the new pile" was cleared.

In 2001, as *Nature* increasingly received submissions electronically, Schön's incoming claims entered the system even faster. From "the new pile," or the equivalent on the journal's New York–based web server, incoming papers were routed to editors with PhD–level training in physics or materials science. The first decision taken by these editors was whether to reject the submissions outright; or, if the research seemed to have far-reaching implications, to send them out for an in-depth review by experts outside *Nature.* For those manuscripts that made the cut—and at least two of Hendrik Schön's submissions did not—editors weighed the peer reviews with responses received from Schön and coauthors, and made a final decision.

The stakes were high. Then, as now, *Nature* and its American competitor *Science* sent out weekly highlights of content to reporters under a news blackout agreement, an embargo. Reporters used the time to understand the sometimes complex science in the papers and warn their own media outlets if a big story was coming up. For some findings, including several of Hendrik Schön's, this system resulted in coordinated media coverage on the days that the embargoes lifted. Research published by *Nature* and *Science* was selected not only to have broad relevance, but, having passed peer review, was also widely understood to have gained the endorsement of respected independent experts.

Schön published data in over twenty different scientific journals, but publication in *Nature* and *Science* brought his claims to a far broader readership. Yet the decision-making process that editors used to accept his manuscripts has never been well understood outside the journals. This is because the review processes, and the identities of reviewers for his papers, have been kept confidential. Anonymity enables editors to solicit honest criticism without worrying that reviewers will hold back for fear of retaliation by disgruntled authors. There is a less clear purpose behind the confidentiality of the content of reviewers' criticisms, which are, in any case, provided to authors. Editors at *Nature* have told me that they are prepared, in principle, to discuss their decisionmaking after papers are published, although in the Schön case, the journal did not do so, and editors have since told me that they have lost many of the records related to their handling of Schön's manuscripts. At *Science,* editors maintain secrecy surrounding peer review even after publication, and even when papers turn out to be fraudulent. At one scientific meeting I attended, the editor-in-chief of *Science* from 2000 to 2008, Donald Kennedy, told the audience that if reviewers ever spoke publicly about their recommendations, the journal would not use them again. Executive editor Monica Bradford explained that confidentiality is a blanket policy, and that editors "don't want to get into 'he said she said'" about why they have made particular editorial decisions. The bottom line is that the path that Hendrik Schön's manuscripts took from "the new piles" to the lifting of the press embargoes has been a black box.

I set out to try to change this, asking for information from reviewers and coauthors of Schön's papers. I ended up with information concerning seventeen papers, twelve of which were eventually published. Although I was not always confident that I had seen the complete file behind editors' decisionmaking, in many cases I did see documents from the time (where I am aware that the information I received reached my source from Schön, I have said so in the account that follows). I am sure that, had the journals decided, following Schön's disgrace, to turn over records relating to the acceptance of his work to a commission of inquiry into the peer review process, a more comprehensive account would have been possible.[4] Still, over a variety of sources and papers, I found that a consistent picture emerged: a sampling of the way in which Schön's breakthrough claims made it into the scientific literature.

NO QUOTA ON DISCOVERIES

The journal *Science* is located in a dark-gray granite block, within walking distance of the halls of the U.S. government in Washington, D.C. It is published by the American Association for the Advancement of Science (AAAS), a lobbying and educational organization that has as its mission "to advance science, and serve society."[5] AAAS describes itself and *Science* as nonprofit, although AAAS tax returns show that advertising in the journal raises more than $10 million each year through the for-profit corporation AAAS Science Publications.[6] The selection process for manuscripts is similar to that at *Nature,* except that the editors' preliminary assessment is conducted with the help of an editorial board of senior scientists.

Even before Schön's blockbuster claims arrived, editors at both *Science* and *Nature* had been keen to publish more physics papers. Karl Ziemelis, *Nature*'s physical sciences editor, told me that from the time he joined *Nature* in the 1990s, the journal began making an effort to publish more physics, in order, he said, to break a widespread perception of the journal as being stronger in the field of biology. "We were flying to conferences and talking to physicists, and our message was, 'we have trained physicists on our staff, and we're interested in what you're doing,'" said Ziemelis. He took on a polite, English butler–like manner as he elaborated on the service ethos according to which, he said, *Nature* tried to run an efficient publication process for science authors with important discoveries to share. The same was true at *Science,* where Brooks Hanson, deputy editor for physical sciences, described if anything a more concerted trend to court physicists than Ziemelis at *Nature.* After 1985, editors at *Science* increased their proportion of physical sciences coverage from 30 percent to 40 percent, in part by soliciting submissions from physicists, and in part by keeping the acceptance rate for physics proportionally somewhat higher than that for biologists, Hanson said. Both journals were interested in solid state and materials physics, fields of research that have implications for the energy, electronics, and computer industries, and that experienced a particular boost following the rush to explore high-temperature superconductivity in 1987. Hanson also pointed to a 1995 *Science* paper that described the phenomenon of Bose-Einstein condensation, in which particles are cooled down until they collapse and form a single quantum entity. The discovery of Bose-Einstein condensation won the 2001 Nobel Prize for Physics.

Schön's manuscripts seemed to be cast in a similar mold. Schön claimed to have discovered important physical phenomena in industrially relevant materials such as plastics. Inevitably, editors at *Nature* and *Science* were in competition to get first crack at evaluating his submissions for publication. For example, Ziemelis, who had worked with organic materials as a graduate student at Cambridge University in the early 1990s and had followed the Bell Labs effort closely for over a year, confirmed that he had felt vexed, sometime in 2000, to notice one of Schön's big results in the competitor journal. Soon after, Ziemelis happened to notice Bertram Batlogg at a scientific conference. Ziemelis went up, said hello, congratulated Batlogg nicely on the paper, and then asked politely why the group had sent the result to *Science*, noting that *Nature* would also have been interested. When I asked Ziemelis whether it had crossed his mind that a series of spectacular submissions coming in from one person might be a sign of fraudulent publication mania, he said no. He had been working with a different picture. As submissions kept coming, editors at *Nature* gathered that Schön, an intense, hard-working postdoc, had spent a year or two building up an apparatus for sputtering his insulating layer of aluminum oxide. Schön could apply an electric field through this insulator to influence the surface of a wide range of materials. Over time, the number of submissions did create an appearance that Schön was very ambitious and was trying to stake as many priority claims as possible, Ziemelis said. But it was hard to blame a young scientist for taking advantage of his hard-earned lead before other laboratories caught up.

Schön's stream of submissions also seemed legitimate to editors at *Science,* especially because of Schön's affiliation with Bell Labs. In the United States, it was well known that physicists at Bell Laboratories didn't face the same administrative and teaching obligations as university researchers, Hanson said. They might therefore be very productive.

As a result, both sets of editors were open, at least in principle, to the possibility of publishing a series of fast-paced physics discoveries from a young Bell Labs researcher. Many of Schön's papers were also groundbreaking enough to fly through the preliminary rounds of assessment. Yet, the odds should still have been against publication. At *Science,* only one fifth of submissions make it through preliminary assessment, and only a quarter of those make it through peer review. "Peer review is a rational information-gathering process," insisted Hanson. But he added that editors keep the same criteria in mind throughout peer review as they do during preliminary assessment. How groundbreaking is this submission? What potential does it have to "advance science"? This means that a typical rejection at *Science* does not occur because a paper is shown to be wrong, Hanson said, but because the review has revealed to editors that it isn't as important as it seemed. "A reviewer may say, 'I am disappointed that this is more incremental than I had hoped when you first asked me to review it,'" said Hanson. On the other hand, Donald Kennedy, *Science*'s opinionated editor-in-chief from 2000 until 2008, has also written effusively of what he called "gems in the mailbag,"[7] sparkling submissions that caught an editor's eye from the first moment of their arrival in the paper-tracking database *Science* used to log thousands of submissions each year.

This was especially relevant to Schön. Journal editors and reviewers didn't know this at the time, but Schön was working backwards. He started from conclusions that he knew would hit other scientists' buttons, and then assembled data to demonstrate those. No wonder that editors at *Nature* and *Science* found many of his papers interesting enough to send out for peer review, and no wonder also that their reviewers tended to agree. Of the evaluations I learned about, all were either positive or expressed an openness to publication of a revised version of the submitted manuscript. Although I heard rumors about negative reviews that journal editors had supposedly ignored and felt an obligation to chase these down, I did not end up with evidence that they had occurred. The reviews I saw or learned about were positive about the significance of Schön's discoveries, if true, as well as about the quality of his compelling data, which was mostly taken to be providing clear evidence for his reported effects. Some reviews were surprisingly brief, given the gravitas that both journals place on their peer-review systems. But others contained thoughtful questions, or sounded a critical tone, raising questions about Schön's methods, data, or data interpretation. The way that reviewers and editors tackled those questions, when working with undeniable enthusiasm for the data on offer, turned out to be a crucial factor in Schön's case of fraud.

"THE SUBJECT IS DEAR TO MY HEART"

A typical example kicked off on Friday, January 5, 2001, when *Science* was the target of an incoming Schön manuscript. The paper described the creation of a Josephson junction, in which two slices of superconductor were sandwiched together by a slice of conductor, enabling a superconducting current to tunnel through from one side to the other. This was an ingenious experiment that another Bell Labs MTS had told Schön would be considered an elegant proof of superconductivity in organic crystals if it succeeded. In his manuscript, Schön wrote that he had used the field effect to tune three adjacent slices of a crystal of carbon-60 into the superconducting-conducting-superconducting sandwich. Preliminary assessment at *Science* usually takes ten days to two weeks, Hanson said, and a week after submission, Teun Klapwijk, a professor of physics at the Delft University of Technology in the Netherlands, opened his email to find a note from Ian Osborne, *Science*'s physics editor in Cambridge in the United Kingdom. Osborne told Klapwijk he was writing about a paper from "Schön et al. (Batlogg)." Schön was still relatively unknown, but by dropping the name of Batlogg, Schön's highly regarded coauthor, Osborne got Klapwijk's attention.

Klapwijk was a superconductivity expert, patiently mannered, with tousled grey hair. He had a special interest in trying to ensure that Schön's work was reported accurately, as several junior researchers at Delft were working to try to build on the findings. At the same time, he was a senior professor who was naturally sympathetic to new developments in superconductivity. "The subject is dear to my heart," Klapwijk told Osborne, as he agreed to do the review.

Science expedited the paper to Klapwijk by the next working day. But when Klapwijk opened it, he realized that he had seen it before. Batlogg had sent it to him several weeks earlier. Although Klapwijk had responded with some critical comments, he could see that these had not been taken account of in the version of the paper at *Science*, and he wrote his review by repeating them. It was—and is—common for scientists to send a paper to colleagues who they expect, or even intend to recommend, to be anonymous peer reviewers, in order to get an idea of what criticisms might come up or, as one scientist put it more cynically, to sweeten them up. Brooks Hanson at *Science* told me that editors want to know who has seen a paper prior to submission and would rule those people out as reviewers. *Science*'s guidelines at the time also stipulated that authors must declare who has "reviewed" their paper. But the comments had not been incorporated as they might have been in a formal review process, and the journal editor neither asked nor was told that Klapwijk had already seen the manuscript.

In his review, Klapwijk urged publication, but also asked (at least) two technical questions. First, he pointed out that Schön had marked up two graphs with "arbitrary units" rather than amperes, the units of current. This meant the graphs only showed the pattern of rises and falls in current that Schön had measured through his devices, not the overall magnitude of the effects. Klapwijk was frustrated that he could not run basic checks on the data and, perplexed, asked to see the actual numbers.

Klapwijk's second question was about the form of the data in one graph, which he could make out even though it was marked in arbitrary units. When two superconductors are separated by a slice of conductor, a graph of current passing through the device is expected to show a certain characteristic shape familiar to Klapwijk as a superconductivity expert. Schön's data looked characteristic of a different device, one in which two superconductors were separated by a slice of insulator. Investigators later concluded that this was because Schön had mistaken the two types of device and so chosen the wrong type of data for his report, which was ultimately retracted.[8] But because Klapwijk assumed that Schön's data were honest, he believed that the data revealed Schön had built and measured a different device from the one he thought and asked that this be double-checked.

Having given a positive recommendation, Klapwijk did not see Schön's paper again until April 13, 2001, when it appeared in *Science.*[9] He reread it, hoping to see that his questions had been answered, but found, to his disappointment, that they hadn't. There were some changes to the paper, it was true. Instead of saying that the material in the filling of the sandwich device was a conductor, Schön and his coauthors said they did not know whether it was an insulator or a conductor and added brightly that "further experiments will be needed." Klapwijk thought the answer superficial. But he was happy to see that in place of arbitrary units, Schön had included exact measurements of current marked in amperes and had run the checks that he would have done during review had the numbers been available. To his surprise, he found that the current Schön reported having measured across the device was unrealistically small; this should have been impossible to measure.[10] Suddenly, a paper that he had recommended enthusiastically for publication was beginning to unravel.

Klapwijk's experience in having recommended a Schön paper, but having reason to feel disappointed about the way his questions were answered, was common to other reviewers. Schön's first *Nature* paper on superconductivity in organic crystals went to print promptly after a reviewer recommended publication in August 2000. "This is admittedly a very fine piece of work which I would gladly see published in *Nature.* I wish to congratulate the authors for their technical achievement . . ." wrote the reviewer, another superconductivity expert. Yet the reviewer also asked that some additional information be included, such as the contact resistance—the resistance of the metal contacts attached to Schön's crystals, which could potentially interfere with the measurements. In the final version of the paper,[11] Schön wrote that he had coaxed the resistance of the material down to one hundredth of its normal state, but marked his graphs in "arbitrary units" and did not reveal the value of his contact resistance.

This failure to answer questions precisely also occurred during review processes at specialist physics journals. In February 2001, Schön sent a paper to the *Journal of Applied Physics* claiming to have measured ballistic transport, in which electrons are able to flow from one point in a material to another, without bouncing around in between. Schön had convinced his coauthor, Batlogg, of the result in his usual staged way. First he floated the idea, telling Batlogg in an email that he had measured "something crazy, like a negative resistance." Months later, he followed up, telling Batlogg that he'd had a "great revelation" that the thing he had seen must have been ballistic transport. A feature of this effect is that as electrons flow from one point on a semiconductor surface to another and overshoot points in between, they can create the illusion of negative resistance. Schön provided a batch of fabricated data that showed the effect.[12] The fabrication wasn't detected by a reviewer of the resulting manuscript, Alberto Morpurgo, who worked at Delft with Klapwijk and informed the editor that "the quality of the data presented is very good and the evidence for ballistic transport is substantial." However, Morpurgo had several questions that he felt should be answered before the work was published. Like Klapwijk, he queried the use of arbitrary units rather than real units. He also asked whether the deposition of aluminum oxide had damaged the surface of the organic crystals. Schön did not respond to these points in the final version of his paper. And while he did respond to some of Morpurgo's other questions, he did so only with his same relentless literalism. For example, Morpurgo had asked for information about the size and shape of the electrical contacts Schön had attached to the crystal surface, saying that he needed the information to try to understand the graphs showing negative resistance in detail. "With small contacts one expects to observe negative resistance also after the first . . . 'peak', but this is not what has been observed," Morpurgo explained. Schön wrote in the paper that "the absence of a negative resistance after the first . . . peak can be ascribed to the large contacts used in the experiment," parroting the review, but not providing the information about the size of the contacts that Morpurgo wanted. By the time Morpurgo saw a revised version of the paper, it was in print.[13]

Why did editors let Schön's papers into print without ensuring that the reviewers' questions were answered? At most journals, editors begin the process of acceptance by sending a conditional acceptance letter; they will publish a paper if revisions are made. Reviewers are also able to recommend that editors "accept with changes," without insisting that they see what those changes are. In both cases, reviewers and editors make the assumption that authors will respond to reviewers' questions in their forthcoming revisions. Putting on the brakes after a conditional accept letter has gone out is never something that comes naturally to editors, who told me that if reviews are positive, and authors seem to have responded in good faith, they might publish without requiring that every change has been made to the reviewers' satisfaction. Most reviewers I interviewed did not remember insisting that they see Schön's manuscripts for a second time, and editors rarely asked them to verify revisions either. Schön, meanwhile, could be as disingenuous as he was dishonest. Although, for most papers, I did not see his responses to reviewers, for two papers for which I did, he seemed to have circulated comments to colleagues suggesting that he was in agreement with reviewers, but then published papers without making changes to reflect these private concessions. The reviewer of one *Nature* paper, who, unusually, did see a paper for a second time before it was printed, pointed this out, warning editors that although Schön had admitted in comments that the reviewer was right about one point, his paper handled the point in the same way as it had prior to the review.

More rigor by editors and reviewers would not necessarily have prevented Schön's papers from being published (although reviewers do sometimes bring fraud to light[14]). But, if only the papers had acknowledged criticisms that came up during review, readers might have more easily overcome the sense of awe conveyed by the compelling data, and scrutinized the results more closely. In addition, the technical details that reviewers asked Schön to include would, if provided, almost certainly have made it easier for other scientists to follow up his work in their own laboratories. That didn't happen. In November 2000, for example, Schön sent a paper to *Nature* in which he described having produced superconductivity in polythiophene, a cheap and manufacturable plastic, by applying the field effect to its surface through his aluminum oxide insulating layer. This was the claim later called the "Plastic Fantastic" in the media. The reviews I saw, circulated by Schön to colleagues, were positive. "This is the first time superconductivity can be stabilized in an organic polymer [a chain-like molecule] and, as such, this discovery deserves publication in *Nature*," wrote one reviewer. A second reviewer agreed: "In my opinion, this paper contains the amount of originality and broadness that warrants publication in *Nature.*" But, the first reviewer added gently, "I still wish to make a few remarks, some of them being comments while others are more mandatory." The second used almost identical language: "However, I still have a few questions and comments, that should be addressed." In a letter circulated by Schön, *Nature* editor Leslie Sage offered what (Sage clarified to me by phone) sounded like a conditional acceptance: *Nature* would be pleased to publish the paper, provided that the reviewers' comments were addressed. But, the letter went on to say, *Nature*'s main editorial concern was that the implications of Schön's results be made clearer to nonspecialist readers. What additional insights into superconductivity did the plastic superconductor reveal? Schön was offered an additional three hundred words above the journal's ordinary word limit to answer this. Technical details that the reviewers wanted, but that might not be interesting to general readers, could be relegated to the captions of Schön's figures. So although the letter asking Schön to take on board reviewers' comments was sent in good faith, the *Nature* editor was, thereafter, much more concerned about the broad appeal of the paper shining through than the technical details. When the paper was printed,[15] in March 2001, it resulted in a sizeable news splash. But while Schön had responded to several points raised during review, he had ignored an important technical question: how had his aluminum oxide had been able to withstand the high electric fields he had applied? In the excitement over the new result, crucial technical information needed to replicate the work was left out.

In other cases, editors and reviewers missed out on chances to ensure that Schön's papers were up front about problems raised by the interpretation of his data. Several of Schön's papers on superconductivity were considered problematic by theoretical physicists. Schon had fit his data to a form of the "BCS" theory of superconductivity (named for the American physicists John Bardeen, Leon Cooper, and Robert Schrieffer) that was conceived to describe superconductivity in the three-dimensional interior of a material. But the superconductivity he reported was supposed to be on the surface of the Bell Labs crystals, where superconducting effects were expected to take a different form. Schön, judging from his data, didn't know that. According to Peter Littlewood, a theoretical physicist at Cambridge University and former head of theoretical physics research at Bell Labs, this issue was not subtle. "The whole of the community would have noticed it instantly," he said.

Reviewers did, but they flagged it to journal editors in only the briefest of ways. The reviewer of Schön's first *Nature* paper[16] on superconductivity told me in an interview that he had felt very surprised by the abrupt drop in resistance supposedly measured by Schön, something that was not typical for superconductivity measured on a surface. But the review, it turned out, commented on the issue only briefly. Reasoning in hindsight, the reviewer said he must have assumed that the editors at *Nature* might not be interested in a technical point, so he had not provided too much detail about it. Over a year later, a different reviewer of Schön's July 2001 superconductivity submission to *Science* was even more brisk, at least according to information Schön circulated to colleagues. The reviewer asked in passing whether Schön's numbers corresponded to superconductivity on the surface or in the interior of his material but did not spell out why that could be a point of concern. The question was one sentence in an otherwise positive review that was only a paragraph long and opened with the declaration, "This is a well-written and entirely believable paper." At this, *Science*'s senior editor Phillip Szuromi had the green light. According to a letter circulated by Schön, Szuromi wrote to Schön eleven days after submission to say that *Science* would move ahead on the basis of only one review, both because of the reviewer's enthusiasm and because of the support of an unnamed member of the editorial board. *Science* would send along comments from a second reviewer if they arrived, Szuromi said. He attached a copyright transfer form and asked that Schön send him electronic files so that the paper could be published "as soon as possible."

Acceptance of a paper on the basis of only one review is a clear violation of *Science*'s published guidelines; the journal promises scientists, reporters, and other authorities that research has been reviewed by at least two experts. When I provided a copy of Szuromi's letter to *Science,* I received a reply from the journal's executive editor, Monica Bradford, explaining that this was a pending accept letter, meaning that, in principle, journal acceptance should still have been conditional on further revisions by Schön. But, she conceded, these were not required. Schön's paper was accepted in sixty-four days, in time to appear back-to-back with a paper of his reporting a similar result that had been submitted seventeen days earlier.[17] Both papers later had to be retracted. Despite the appearance of a special exception in this case, two papers by Schön got into *Science* even more quickly than this one. The average time it took for Schön's papers to get into print in *Science* was 83 days, compared with an average of 112 days for the journal as a whole.[18]

At *Nature*, Schön's average time to publication was 121 days. Judging by the papers I learned about, *Nature* had a slightly less enthusiastic response to Schön manuscripts, with Schön's hesitancy to commit to the interpretation of his data apparently counting against him more at a newsy journal that put its priority on accessibility. At *Science*, editors relied on the heavyweight expertise of an editorial board of senior scientists who were impressed by the data alone. On February 5, 2002, Schön sent off two papers, one each to *Nature* and *Science*, on spintronics, in which information is stored in the magnetism or "spin" of electrons. *Nature* rejected the submission without peer review because the paper's implications seemed vague to the editor who received it. *Science* sent the paper out for review, with reviews that one coauthor told me were the most enthusiastic he had ever seen, and the journal expedited the paper into print via *Science Express*, a mechanism for online publication that *Science* had established to allow scientists "to connect with these hot results immediately."[19]

Only toward the end of Schön's trajectory did the enthusiasm of editors and reviewers begin to ease off. Thomas Palstra, a professor of solid state chemistry at the University of Groningen in the Netherlands who had worked as a postdoc with Bertram Batlogg at Bell Labs, reviewed a Schön paper for *Nature* in the spring of 2002. In common with one other reviewer I learned of, Palstra was, by then, beginning to feel concerned about the lack of replicability in Schön's work. He said that he agreed to do the review only on the condition that his comments, which included demands for additional technical information, would be taken seriously. Having insisted, Palstra said he did find the editor responsive to his concerns. "When you review for *Nature* and *Science*, you need to be careful. You need to nag about the details that are forgotten with all the big implications. As an expert reviewer, my first question is, 'is this correct?' If I get notified by *Nature*, I already know it's important. They need my opinion whether it's correct," Palstra said. But Palstra said that when he saw the Schön paper he had reviewed for a second time, some of the technical information that he had requested was included, while Schön and his coauthors argued that information they did not have could be the subject of future work. Palstra agreed.[20]

Speaking in general terms, editors at both journals emphasized that while they try to run thorough review processes, they do not aim to guarantee that everything they publish is right. "We don't want to publish something that turns out to be wrong very soon after it is published," said Ziemelis at *Nature*, adding that if things turned out to be wrong later, *Nature* would accept that as a consequence of the risk associated with publishing frontier research. Brooks Hanson at *Science* said, "Sometimes very preliminary results can be thought-provoking, and encourage new science. Our goal is the advancement of science. Does it advance science? Even a paper that's wrong can encourage new science." The future course of science, not the journal review process is the ultimate arbiter of truth or falsity of scientific claims.

This idea provided much of the rationalization for editors and reviewers to accept publications that they knew contained gaps, on the basis that the results seemed important, if they were true. This raises a pressing question. Given that, for most of his trajectory, Hendrik Schön was riding a wave of enthusiasm into print and getting away without documenting the answers to some of the reviewers' questions, what happened when he was obliged to respond in more detail?

A SIGN OF INTENT

While reporting on possible fraud in the scientific community, the strangest call I have ever received came on a Saturday night in December 2005. My cell phone rang, and when I picked up, a voice said, "You're making trouble for my boss."

I scrambled for a pen. "Who is this?"

The caller did not want to give his name, but my cell phone showed a South Korean number, and I figured that his boss was the cloning researcher, Woo Suk Hwang, to whom I had sent an email request for comment. By that time, Hwang's claims to have cloned the first human embryonic stem cells were in disarray. They began to unravel in November 2005, when Hwang admitted he had used eggs donated by researchers in his lab for his experiments, contradicting assurances published as part of his 2005 paper in *Science*. Then it emerged that producers at the Korean television station MBC had planned to screen a documentary claiming that Hwang's data were fraudulent. Finally, anonymous bloggers started posting evidence of fabrication on the message boards of the Biological Research Information Center at Pohang

University of Science and Technology, showing how images of Hwang's supposed human embryonic stem cells had been stitched together from photographs of different cells.

My caller did not want to discuss any of this. He was only interested in responding to a question I had mentioned in my email to Hwang. The question had focused on apparently duplicated images in a minor paper published in the journal *Molecular Reproduction and Development*[21] about the cloning of pig embryos. The caller defended the work and questioned my ability to understand it. "You have never done this kind of experiment," he said, which was true. I asked him to try to explain the duplication, took down his number, and over the next day, exchanged a couple of further calls and emails with him. He had access to, and was using, Woo Suk Hwang's email address. Eventually, he identified himself as Sung Keun Kang, a man sometimes called Hwang's "lieutenant" and who, by the following May, was co-indicted with Hwang for fraud. Over the next couple of days, Kang and I went back and forth. His position regarding the irregularity I asked about changed twice in response to questions, but eventually he outlined a fairly convoluted experimental method that could explain it. When I ran the method by other stem cell experts, they found it unlikely, but there was no way to prove things had not been done this way. Still, because Kang's position had seemed to evolve, I had the feeling that the discussion was helping him to thrash out a response to a question, rather than getting to the bottom of how the experiments had been done. Kang noted at one point that he had not conducted them himself.

Given that researchers in Woo Suk Hwang's lab reacted with alacrity when a new question threatened to surface in the press, how much more effective were their responses to questions from other scientists? In 2004, Woo Suk Hwang's first breakthrough paper arrived at the journal *Science.* In this manuscript, Hwang claimed to have isolated and cloned human stem cells that were able to develop into any tissue in the body, raising the prospect that patients with degenerative disease might one day be able to provide doctors with a small amount of bone marrow and then have tissue grown to repair their damaged hearts, brains or spinal cords, or whatever was needed. Publication of this result by the prestigious American journal shot Hwang to international stardom.

When it turned out, in late 2005, that the results were fake, the media focus turned to *Science*, and editors requested a review of their process by a panel chaired by John Brauman, a Stanford University chemist and member of the journal's editorial board. Brauman's report described *Science* (and *Nature*) as having reached "special status" in the scientific community. It went on to argue that editors at *Science* could no longer assume that scientists submitting to the journal always did so honestly. In the case of Hwang, the report noted that several pieces of important information only arrived at *Science* after the original submissions. In one dramatic example, reviewers had pointed out that Hwang's data might be explicable if the group had stumbled across parthenogenesis, rather than human cloning. Parthenogenesis, in which two eggs merge to form a functional embryo, is sometimes called "virgin birth" and has been seen in mice, but never normally appears in humans. The reviewers' concern was that as Hwang's workers worked with human eggs in a flask, trying to provoke cloning, they might have inadvertently have triggered parthenogenesis, and then been misled into thinking they had made a clone. Independent experimentalists tested Hwang's cell lines in 2007 and found that they were derived from parthenogenesis, as the reviewers had suspected.[22] But when editors at *Science* forwarded questions on this possibility to Hwang, they received a response saying that the group had considered this and ruled it out. Hwang and coauthors also added a small amount of DNA fingerprint data that seemed to show that the embryonic stem cells they had created had to be derived from a single egg.[23] The reviewers "acquiesced"[24] and *Science* published the paper. Only later was it pointed out, by scientists at the U.S. cloning company Advanced Cell Technology, that the DNA fingerprint data did not look consistent with the instrument supposedly used to acquire it.[25] The paper was subsequently found by investigators to contain fabricated data and was retracted. Today the consensus is that Hwang's group had created the first human parthenogenetic stem cell line—a breakthrough, but not the one Hwang was hoping for. Even though *Science* had run a peer review process that was thorough, and that reached the right scientific answer, the journal still printed a paper that claimed something else. Not only is there no guarantee that a thorough review process will detect a false claim, but even more disturbingly, a thorough review may do little more than reveal to authors what changes they need to make in order to turn a false claim into a more plausible scam.

As it turns out, the possibility that fraudulent scientists respond to reviews by changing their data is well known to professional investigators of suspect science. In 2005, Eric Poehlman of the University of Vermont became the first U.S. scientist to be jailed for scientific fraud. Poehlman had faked studies supporting the hypothesis that menopause has a marked effect on women's health. In 2007, I saw two officials involved in the prosecution speak at a conference on data fabrication. One was a former U.S. attorney for the district of Vermont, Stephen Kelly, who described what he had thought when he read the University of Vermont's investigation report on Poehlman's alleged misconduct. As panel members had tried to understand why data entries in Poehlman's spreadsheets changed over time, Poehlman had provided more than twelve different excuses, ranging from the accidental insertion of modeled data, to data manipulation by someone else using his email account, to piecemeal corruption of his data in a computer crash.[26] Kelly said that as a U.S. attorney he did not grasp all of the science involved in the health effects of menopause, but that the phenomenon of numerous changing explanations for the same evidence was something he recognized from his experience of prosecuting serious crime. "That's something we see in criminal cases," he said. The other official I saw was John Dahlberg, an expert in data forensics at the U.S. Department of Health and Human Services' Office of Research Integrity (ORI), who described how his office had been able to add to the evidence collected by the University of Vermont from copies of Poehlman's applications for funding. In one application, Poehlman had stated that the average increase in interabdominal fat for menopausal women not taking hormone replacement therapy was as high as 17 percent, over just twenty-six weeks. A grant reviewer found this number implausibly high, and Poehlman didn't get the grant. In a later application, Poehlman changed the figure to 4 percent, so that it fell into the range that a reviewer would think reasonable. This time, he got the money. Further investigation revealed more problems; although the study that Poehlman cited was real, the data were not available in a form he could have used at the time he wrote his paper. This helped to explain why the Vermont professor might not have felt too bad about changing the data: they weren't real to begin with. Dahlberg described how the evolution of the data in response to a question had provided evidence not only for careless misrepresentation of results, but for the possibility that the misrepresentations were calculated to convince others. "When a scientist changes data in response to a reviewer's comment, that's a sign of intent," Dahlberg said.[27]

On hearing about these shenanigans by Eric Poehlman and Woo Suk Hwang, it's easy to ask why reviewers or editors aren't more suspicious. Why don't they realize that with a publication in *Science* or a major government grant at stake, some scientists might be prepared to give reviewers and editors whatever data they want to get through? Part of the reason, reviewers told me, is that honest scientists sometimes also come back with different data after a round of criticism. If a reviewer has spotted a real problem in the experimental design or analysis, it is legitimate for a scientist to repeat the work to try to remedy the problem. Reviewers of publications often said they felt their job was to ask questions and seek clarifications, but that once they were presented with clean data from a well-described experiment, they felt a responsibility as scientists to allow that evidence to reach a wider audience, even if they personally were not convinced that the scientific claim was correct.

Seeing how easily fraudsters can take advantage of this openmindedness, using criticism to hone their fraud, it's tempting to ask if there is anything positive about the thorough review of a fraudulent paper. The answer is probably yes. In 2005, as *Science* began to prepare Woo Suk Hwang's paper on human embryonic stem cells for publication, editors asked Hwang to provide higher resolution photographs of some of the stem cells. In response, *Science* received images that included distorted, miscaptioned duplicates of photos shown elsewhere in the paper. The editors did not notice and printed the duplications. But within weeks, readers of Hwang's work had noticed the problem and began to take a closer look, finding further irregularities in the large amount of published data that Hwang had been compelled to submit to *Science* to get his claims published. Apparently, there was a trade-off. Hwang was able to get published in *Science* because he or his lab members were prepared to change their data, partly in response to reviewers. But the cloud had a silver lining: the more information the group provided, the greater the chance that some of the documentation would eventually be publicly questioned.

"WE'D LIKE TO THANK THE REFEREE FOR THEIR HELPFUL COMMENTS"

The acceptance of Hendrik Schön's first few papers had been helped by his coauthorship with Bertram Batlogg, a senior scientist known to reviewers and editors, who was also a skilled communicator. But in 2000 Batlogg moved to the Swiss Federal Institute of Technology, and by 2001 most of Schön's submissions no longer had Batlogg as a coauthor.

Schön had learned a few things by then. He knew it was important to be friendly toward reviewers, and to thank them generously for their valuable comments. He was not too shy to solicit editors politely for their support. But he had a major problem. Schön sometimes faked data without understanding the physical concepts involved, and so risked producing data that puzzled reviewers.

The most significant example occurred in May 2001, when Schön sent a paper to *Nature* in which he claimed to have turned a single layer of organic molecules into the channel of a transistor, an electronic switch. Schön was the only physicist on the paper, which was coauthored with a Bell Labs chemist, Zhenan Bao. Because the molecules arranged themselves spontaneously, Schön and Bao called the device a self-assembling monolayer field-effect transistor, or SAMFET. Leveraging the claim that the channel of the transistor was only one molecule thick, Schön and Bao described the SAMFET as an important step toward "molecular-scale electronics," in which computer chips would be built and patterned at scales hundreds of times smaller than is possible with today's silicon-based technology, setting the stage for future leaps forward in computing power. To support the conclusion that the transistor worked, Schön duplicated some of the same transistor curves and circuit outputs that he had used in two earlier papers on organic crystals.[28] Those curves had worked out well for him on the earlier occasions.

Nature was interested. The editor, Liesbeth Venema, said that she would send the paper out for review. But when the reviews came back, several weeks later, there was a mixture of good and not-so-good news for Schön. The first reviewer told me that he had recommended publication, but also demanded more information about the details of the method used to make the devices. He also said he had emphasized that he wouldn't have expected to see transistor data like that Schön showed. Schön had designed the SAMFET so that one part of the device, the channel, was on a tiny scale, while other parts, such as the gate insulator, were much larger. This meant that the voltage applied to the gate was too far away for the molecules in the channel to be strongly influenced by it, the reviewer said. It was as if Schön claimed to have moved iron filings on a table by holding a magnet underneath but with the tabletop being so thick that the magnet's influence could not have penetrated it. Even though Schön had not mentioned that the data in his paper were taken from his earlier papers on more conventional transistors, the reviewer had unknowingly hit on a consequence of the fabrication.

The second reviewer was skeptical too. He questioned some of the numbers Schön had reported in the manuscript and asked whether Schön had obtained consistent results from other organic molecules. He also challenged the propriety of Schön's claim to have made "molecular-scale" devices. While it was good to have measured transistor characteristics on a single layer of molecules, the channel in the SAMFET contained thousands of molecules, not just one or two, the reviewer felt. Even for someone who believed the data, the claim about "molecular-scale electronics" seemed overblown.

These were tough reviews, but they were not rejections, and the questions only meant that Schön had to do more work. Within a few weeks, he and Zhenan Bao had worked up a set of responses and a revised paper that the reviewers agreed, with some reservations, could be published. According to information circulated by Schön, he added information to his paper at this stage showing transistor data supposedly acquired from a second SAMFET, made from a second organic molecule. The new data were not too hard to generate; Schön took the data that he had produced for his first SAMFET, and manipulated and recaptioned them as coming from a second.[29] He also included information about as many as six molecules in a table that he added to the manuscript. Unbeknownst to Schön's coauthors, the editor, or the reviewers, the review process had done little more than encourage Schön to engage in further fabrication. Schön thanked the reviewers for their valuable comments, which had very much improved the quality of his manuscript.

Nature accepted the paper for publication 167 days after submission. But before Schön could bask in the excitement of this accepetance, he had to do more work. In the course of review, he had apparently shared with *Nature* news of promising, preliminary results showing that, in future, it might possible to reduce the number of active molecules in the SAMFET down to one or two, which would help to justify the language that a reviewer had regarded as somewhat overblown. These data were not included in the manuscript, but Schön followed through on the promise with a submission to *Science*, eagerly citing the paper under submission at *Nature*, and so playing the credibility of one journal off against the other. A reviewer for *Science* told me that he had also raised questions about the puzzling-looking transistor characteristics, but nevertheless recommended publication. *Science* sent the paper into print without bouncing it back to the reviewer for a second look.

Both papers made a splash. The claim in *Nature* to have made progress toward "molecular-scale" computing hit the headlines in numerous media outlets. The claim in *Science* to have made a single molecule transistor was later named as part of the journal's "Breakthrough of the Year" news coverage.[30] Although Schön had not been able to explain why his data looked the way they did, he had been able to get them published by fabricating more of them. Even if the results were unexpected, the generous supply of data suggested that Schön must be measuring a real physical effect of some kind. But the back and forth in the review process at *Nature*, and the gruff suggestion that Schön was overselling his data, had pushed Schön to a new level of excess.

The data that Schön published in the fall of 2001 involved at least three duplications. Although these weren't detected during review, they were going to come back and haunt him.

In the meantime, Schön's habit of avoiding technical question also began to catch up with him. In October 2001 he received a visit at Murray Hill from a Stanford colleague, David Goldhaber-Gordon, who had been reading some of the papers on organic electronics and wanted to offer some advice. Goldhaber-Gordon told Schön that many outside scientists were beginning to tire of the results because the papers did not explain enough about how he had achieved them. He suggested that Schön might want to write a long technical paper that focused on his method for sputtering aluminum oxide. This was the issue that several reviewers of earlier papers on organic electronics had asked about, but that Schön had not addressed. Goldhaber-Gordon did not try to pressure Schön, but he seemed genuinely concerned for a talented colleague who was suffering reputation damage because he hadn't realized how important it was for a scientist to share his methods, as well as his results.

The idea that Schön should write a long technical paper was apparently also discussed with him by others, including his managers at Bell Labs and his collaborators Bertram Batlogg and Christian Kloc. But it was the gentle request of an outside colleague that triggered him to action, Schön told Goldhaber-Gordon at the time. He wrote in an email that following this visit, he stayed up through two nights and one day of writing to produce the first draft of a paper about sputtering. There was a lot of data in the paper that later turned out to have been fabricated,[31] but on a first read-through it gave the impression that Schön had spent many months sputtering, painstakingly depositing aluminum oxide that could withstand the high fields needed for his field-effect experiments.

The result was a paper that was different from those that Schön had published before. It was not a hot manuscript of the kind that he had sent to *Nature* and *Science*, but contained technical information that would not be likely to interest the wider scientific community, let alone the general public, but that might, on the other hand, be very useful to a small band of specialist researchers—the people who were best placed to transform his exciting new data into the stepping stones of scientific progress. If only *Nature*, *Science*, or even the more specialist physics journals had insisted on methodological details being included in the breakthrough papers they had published, others reading Schön's work might have tested the validity of his work in their own laboratories more quickly. The publication of technical details would also have helped secure Schön's reputation as a solid contributor to the scientific community. But Schön was not thinking along these lines. Over two years, he had so cultivated his skill and interest in flashy publications that he barely knew how to publish something of a more technical nature and, when he sent his draft back to his outside colleague for comments, he could not help asking for an additional word of advice. "Do you have any idea," he asked, "what kind of journal might publish such a manuscript?"

7

SCIENTISTS ASTRAY

Joe Orenstein later remembered that he had first learned about the Bell Labs work on organic molecular crystals from *Science* magazine. Unable to sleep, the former Bell Labs member of technical staff had put on a bathrobe and gone into the kitchen in search of some crackers. He found himself flipping through a copy of *Science* that was sitting on the kitchen countertop. One report caught his eye. It was a paper coauthored by Bertram Batlogg about an organic crystal putting out light.

It was mid-2000, and Orenstein was a professor at the University of California, Berkeley. Seeing the paper reminded him of his time at Bell Labs, and of Batlogg, his old colleague, who had seemed to get caught up in managerial and administrative responsibilities over the years. He was amazed to see Batlogg shaking that off and going back to doing science—and what an incredible piece of work it was too. "I thought, 'There goes Bertram. The cream always rises to the top,'" Orenstein said.

Orenstein's lab had also had some good results that year, and he had been traveling to conferences regularly to present them. He saw Batlogg talking about organic crystals at many of the conferences. The talks always had a major impact and were very interesting to Orenstein, who specialized in using laser light to probe the fundamental properties of quantum materials, in which electrons are strongly influenced by each other.

Orenstein also noticed that the Bell Labs group seemed to have processed their crystals heavily. They were making field-effect transistors by evaporating metal contacts, depositing a layer of aluminum oxide insulator and a gate electrode in order to draw charges to the surface of the samples. The processing steps could be expected to introduce artifacts into the data or complicate the interpretation of what was really going on. At one of the conferences, Orenstein went up to Batlogg and asked whether he could have some of the group's unprocessed crystals to measure in his own lab. Orenstein's optical techniques would enable him to probe the electronic structure of the materials accurately without attaching metal contacts or depositing an insulator, he explained. Through a joint affiliation with Lawrence Berkeley National Lab, he also had access to the Advanced Light Source, one of the brightest sources of infrared light in the world.

To Orenstein's surprise, Batlogg took time out of his hectic schedule to talk through the idea and arranged from Zurich for Christian Kloc to send some of the Bell Labs samples. By March 2001, a package of pretty yellow-orange tetracene crystals arrived in Berkeley, California. Orenstein and his graduate student Chris Weber started out making preliminary measurements with an infrared lamp in their lab at Lawrence Berkeley, intending to progress onto the more powerful ALS if these were a success. They put a tetracene crystal into a cryostat and lowered the temperature, as was necessary for making clean optical measurements. As it took at least fifteen minutes for the cryostat to cool, they went to do something else, with the intention of coming back later to start the measurements. Their cryostat contained two windows where the infrared light could go in and out, but when Orenstein and Weber came back and looked through a window, they couldn't see the Bell sample. It wasn't in the sample holder where they had left it.

"We said, 'oh, it's fallen off,'" said Orenstein. Feeling a bit foolish, they opened up the cryostat to have a look. But the crystal hadn't fallen anywhere. It had completely vanished, leaving nothing but a fine residue of fairy dust.

In a follow-up test, Orenstein and Weber stayed in the lab and looked through the windows while they cooled the cryostat with a second tetracene crystal inside. Sure enough, when the temperature got down to about 100 degrees above absolute zero, the Bell sample shattered into many tiny shards. They repeated the experiment several times and lost several samples. Knowing that organic crystals were held together by very weak van der Waals forces, Orenstein figured that the delicate materials couldn't take the strain of being cooled down while being gripped tightly by the sample holder. They tried using rubber cement, a soft glue that could hold the crystals in place more gently, and tried again. Again they found that the samples cracked or self-destructed when they were cooled. Orenstein said he felt dismayed. He understood that in similar crystals, researchers working with Batlogg had been able to measure such spectacular effects as the fractional quantum Hall effect and superconductivity at only a few degrees above absolute zero, even after the additional duress of depositing metal contacts and a layer of insulator. He couldn't even put the samples into a cryostat and cool them to 100 degrees without breaking them. "I felt like a clumsy oaf," he said.

When Orenstein contacted Batlogg about the problem, Batlogg was not inclined to attribute the problem to differential skill. Instead, his suggestion was that maybe, with all the demand for tetracene and pentacene created by different research groups trying to follow up on Bell Labs' results, suppliers had begun to compromise the quality of the chemicals they sent out. The crystals that Kloc had sent to Orenstein looked like the samples measured by Batlogg's former postdoc, Hendrik Schön, but perhaps they weren't as high in quality, Batlogg suggested.

Orenstein regarded this explanation as "tangible." He and Weber had to settle for measurements of the Bell samples made at higher temperatures than was ideal. They went on to get some reproducible results, suggesting the crystals were very consistent in quality. But because the more remarkable effects reported by the group at Bell Labs had only been seen at very low temperature, the new data didn't add many insights to the cutting edge of the field. Orenstein and Weber wrote bluntly in a report for the Advanced Light Source at Lawrence Berkeley National Lab: "these results . . . do not put us particularly close to our scientific goals."[1]

Despite the disappointment, Orenstein accepted Batlogg's explanation, and certainly didn't consider the possibility that there was a problem with the Bell Labs work. When Orenstein had worked at Bell Labs as a young scientist in the 1980s, it had been easier to get a permanent job, *Nature* and *Science* papers hadn't been in much demand, and the idea that anyone might want to report a spectacular breakthrough in anything other than a careful and accurate way had been unthinkable. Orenstein also wasn't trying to replicate the Bell work so much as to take it as an inspiration for further studies. Orenstein continued to talk with at least two other research groups in the hope of getting organic crystals that would be strong enough to stay in one piece when cooled down, but they never overcame the problem with tetracene. Chris Weber spent around eighteen months working at least half-time on the topic, entirely because of the Bell Labs claims. Orenstein reasoned that it might be possible to make progress with a bigger operation than just his one lab. Suppose not just one or two, but maybe as many as a dozen students worked to follow up on the Bell Labs result? Surely they would get somewhere interesting.

MINNEAPOLIS

Two thousand miles to the east, twenty-odd blocks of the University of Minnesota's Minneapolis campus stood in the crook of a dull gray river, an unlikely looking stretch of the Mississippi. In 1999, Dan Frisbie, an assistant professor of Chemical Engineering, was building up his lab in the campus's Amundson Hall. Frisbie got to know Hendrik Schön early. "Before he began his publishing rampage," Frisbie later said. He remembered having noticed the young German in Santa Barbara, California, at a meeting of the Electronic Materials Conference in June 1999. Schön reported measuring mobilities of charge in pentacene that increased dramatically at lower temperatures in the way previously reported by the organic crystals pioneer Norbert Karl. If anything, Schön's mobilities were higher. Frisbie was speechless. He looked around the small session to take stock. There were about thirty people in the room. "But nobody stood up and said, 'wow, this is truly amazing,'" said Frisbie.

Frisbie saw Schön again a few months later, at the Materials Research Society Fall meeting in Boston, where Frisbie had grown up and earned his PhD before moving out to Minnesota to take his job as a professor. This time, he sat down next to Schön in the session on organic electronics, and the two men made a comfortable connection. Schön, about three years younger than Frisbie, seemed shy but easy to talk to. In the session break, Schön told Frisbie that he had seen ambipolar transport, in which the current is carried by either positive or negative charges, in pentacene. "I thought this was a terrific result. I congratulated him on the spot," said Frisbie. Within two months, Schön had published a paper in *Science* that described how he had done it. Schön had built a field-effect transistor in which he induced positive or negative charges on the surface of an organic crystal on demand through an insulating layer of aluminum oxide.

The next time Frisbie saw Schön, the organic crystal field-effect transistor was beginning to shake things up in earnest. In 2000, the annual March meeting of the American Physical Society came to Minneapolis. Again, Frisbie sat next to Schön in the organics session. When Schön saw Frisbie, he took out a preprint about the trapping of charge in organic films, an area in which Frisbie had expertise, and gave it to Frisbie. Then Schön got up and presented a string of incredible results that included superconductivity and the quantum Hall effect. Schön looked anxious, stuttering a bit on the early slides, but then got into his stride. After the talk, he came back to his seat and Frisbie offered his congratulations. Schön said thanks. He admitted feeling nervous. Frisbie tried to imagine what it would be like to stand up and present such an incredible series of results.[2]

Back at the lab, Frisbie decided to give a graduate student, Reid Chesterfield, the task of replicating some of Schön's work. If his lab could figure out how to apply the field effect to the surface of organic crystals, they could try out a variety of experiments and take things further. Chesterfield also wouldn't have to work alone. In the physics department at Minnesota was Allen Goldman, one of the world experts in superconductivity. Goldman's postdoc, Anand Bhattacharya, was raring to use the field effect to try to induce superconductivity on the surface of organic crystals. Toward the middle or end of 2000, Frisbie and Goldman's labs geared up to what Bhattacharya called "a double-barreled effort" to be the first to replicate, and even to surpass, Bell Labs' promising lead.

Like Frisbie, Allen Goldman hadn't jumped onto the organic crystals bandwagon straight away. At the end of 1999, he and Bhattacharya were still discussing applying the field effect to ultrathin films of other materials. Goldman wanted to study the onset of superconductivity on the surface of a material. But the films his lab used to study this had to be deposited and measured using heavy specialist equipment. The Bell Labs field-effect transistors seemed a potentially more convenient experimental set-up. Bhattacharya and Goldman were caught by the idea at the March meeting in Minneapolis. Schön spoke in a small session on a Tuesday. On Wednesday, Goldman went to a plenary session talk given by Bertram Batlogg, who had arranged all the new results with impeccable scientific logic. Goldman described to Bhattacharya later how, prompted by the scientific logic of the talk, he spoke aloud in the audience at one point: "I wonder if this is superconducting," he said. Then Girsh Blumberg, an MTS in the Bell Labs contingent, leaned over and said, in a dramatic stage whisper, "it is superconducting." There were some questions circulating about how the devices had been made, but there was also a kind of swagger to the community, as if they were going to a new frontier and prepared to try their luck even without those questions having been answered, Bhattacharya remembered. "Sometimes you suspend critical judgment when you see possibilities opening up."

The first step was to build a field-effect transistor setup at Minnesota. Chesterfield and Bhattacharya worked closely from the Bell Labs papers. Following Christian Kloc's publications, Chesterfield put together a crystal growth apparatus in Dan Frisbie's lab and taught himself to grow organic crystals. He also learned to attach metal contacts, and made some basic measurements of the mobility of charge. He found his measurements varied from one crystal to the next, and never measured mobilities higher than about a tenth of those Hendrik Schön had reported, but in occasional samples of tetracene he sometimes felt he saw a leap in current, the evidence for trap-free charge transport that Schön had reported in late 1998. Still, this was only a precursor to the finding they wanted. "The killer result was the FET," Chesterfield said, the field-effect transistor.

The labs developed a routine. Chesterfield grew the crystals and put metallic contacts on, then gave them to Bhattacharya for the deposition of aluminum oxide. If all went well, Chesterfield could then add the gate electrode, and they could start making field-effect measurements. But they often never got that far. Frisbie remembered Chesterfield showing him devices in which the aluminum oxide layer had buckled, ripped, or failed to stick. The surface of the crystals was frequently damaged and far from the clean, smooth surface that would be ideal for field-effect measurements. One time, when Chesterfield opened up the sputtering machine after a round of sputtering, he found it empty, the fragile crystal having disintegrated away into nothing, apparently after being peppered at too high a power with particles of aluminum oxide. Bhattacharya said they made progress on a steep learning curve when it came to growing crystals, but that when it came to putting down the insulator, "all bets were off."

After around six months to a year of failure, Bhattacharya decided to write to Schön directly. He did not write sooner, maybe because Schön was a bit of a celebrity, he said. But Schön was receptive, friendly, and encouraging. He urged the Minnesota group not to give up. He promised them that if only they were persistent, they would succeed in replicating his work. He offered to answer any questions they had.

They had many. They asked how far Schön had placed his sputtering target from his sample. He said fifteen centimeters. They placed their target fifteen centimeters from their sample too. They asked for the wattage he was using to fire the gun inside the sputtering machine and made sure that theirs matched. When Schön said that he sometimes repaired damage caused by sputtering the crystals by placing them in hydrogen, Chesterfield plumbed a hydrogen supply so that they could do the same in Minnesota. The group even asked which vendor Schön used to buy aluminum oxide. He gave the name of a vendor different from theirs. Although they didn't expect this to make a big difference, it was easy to change, so they did. When nothing worked, they double-checked how far the sputtering gun was from the target samples. This time, Schön said about four centimeters.[3] This didn't match what he had said before, so they figured that maybe he had changed what he was doing. They changed too. Chesterfield let out a heavy sigh as he recounted all the different things that Schön suggested and that they tried. "None of it worked." Even when the group managed to get a layer of aluminum oxide onto a crystal without causing obvious damage to the crystal—which was rare—they still couldn't apply a high voltage to the gate electrode without the insulator rupturing and short-circuiting out the device before field-effect measurements could begin. "I said, 'please,'" said Bhattacharya, sounding as if he was repeating a prayer. "Please, just get me to the field-effect transistor. I'll work from there."

In December 2000, Bertram Batlogg came to give a colloquium on organic crystals at the University of Minnesota, "and we cornered him," said Bhattacharya. The two young researchers described their failures to Batlogg. Batlogg seemed concerned and spent more than an hour talking everything through. Bhattacharya said he had not felt Batlogg was in touch with the experimental details but that it had been good to talk to him. Chesterfield also said he came away from the discussion feeling re-energized, although he also did not remember specific suggestions for new things to try. Looking back, he said, "I think you clutch at straws when you've been trying for so long."

Things weren't looking good, but as a professor trying to build up a new lab, Frisbie was persistent. Still, eventually, he and Chesterfield decided to cut their losses and started a new project working with a thin organic film that had nothing to do with Hendrik Schön's technique. They began to get some results straightaway and later measured ambipolarity, the one of Schön's results that had originally impressed Frisbie, in a thin-film transistor made with terthiophene, a molecule containing three rings of carbon and sulphur atoms.[4] Allen Goldman explained to Bhattacharya, who had committed himself to jumping into the new field, that it was important not to put all his eggs in one basket when doing research, and they went back to working with thin films of other materials. Bhattacharya surfaced to notice that his contemporaries were getting papers published and were being offered jobs as junior professors at good universities, and that he wasn't. His postdoc research, expected to be a short career stage lasting two or three years, went on to last five.

By the time news came out that Schön's work was fraudulent, Chesterfield said he hardly read it. The information filtered through to him only gradually, especially in 2003 when he met another researcher who had tried to replicate Schön's work, Vitaly Podzorov of Rutgers University in New Jersey, at a conference. After talking with Podzorov, Chesterfield went home, took one of his old crystals out, and made some final measurements. "Then I concluded Schön was lying," he said. In total, he had spent two years trying to replicate the Bell Labs results. But he hesitated to say that it was a complete waste of time. As a graduate student, Chesterfield had learned how to grow organic crystals, how to deposit contacts, and how to measure the resistance of contacts. He said he learned a lot about transistors by failing to make one.

Bhattacharya also had mixed feelings. He said that he learned a lot about how to do science, how to follow his nose, and not go after things in a "holy grail kind of way," he said. He had a good introduction to the hitches and surprises that could ruin a field-effect experiment, as they nearly all happened. In 2001, Anand Bhattacharya was joined by a graduate student, Melissa Eblen (now Eblen-Zayas). When they didn't get aluminum oxide to work as an effective insulator for organic crystals, they tried another material, strontium titanate. That did not work either. But eventually, two other researchers in Allen Goldman's group, Kevin Parendo and Sarwa Tan, worked to produce a field-effect transistor set-up that used the strontium titanate developed by Bhattacharya and Eblen in trying to follow up on Schön's work, and combined it with a thin film of bismuth, deposited in Goldman's heavy specialist equipment, to make measurements of field-effect induced superconductivity.[5] "I will never experience so much satisfaction again in my whole life," said Bhattacharya.

But it didn't make up for the wasted time. When I visited Allen Goldman's lab four years after the end of the Schön affair, he printed out a twenty-eight page review paper that he, Frisbie and twelve others had written as a result of a U.S. Department of Energy workshop on using the field effect to modify the electronic properties of materials. The paper made the case that there was still plenty of other interesting science going on in the field in which Schön had seemed to be a pioneer.[6] Downstairs in his lab, we passed a small cabinet, which Goldman thumped. The cabinet rattled as if it were full of skeletons, or old samples. "Organics graveyard," said Goldman.

Dan Frisbie also sounded disappointed. Despite the major attention given to the exciting claims at meetings and in journals, it had taken a long time to realize that Schön's data were invalid. Although other scientists were having difficulty replicating the work, they did not report their failures publicly. No one claimed success, but it had seemed possible to Frisbie that people were keeping the details of their progress close to their chest, at least until they were ready to publish. It had been easy to feel isolated following up something that everyone seemed to believe, but that wasn't true. "Can I ask you something," said Chesterfield when I interviewed him. "Did anyone try as hard as we did, and still not get it?"

Frisbie said that if something like this happened again, he would pick up the phone and make calls to other labs to try to get information circulating. At the time he was more junior, less confident, and didn't make the calls. Things had only started to come into perspective by chance. In 2001, Frisbie happened to be invited to a wedding in Greece. There, he later remembered, he had met another researcher who was working to follow up the Bell Labs claims in Delft in the Netherlands. Chatting against the noisy background of a large outdoor Greek wedding reception, Frisbie began to realize that people were having trouble following up Schön's results elsewhere.

FROM DELFT TO SENDAI

Four thousand miles east of Minnesota, cycle lanes meander between housing estates and canal-front homes to bring staff and students into labs at the Delft University of Technology. Over the road from a low-lying willowy pond, a group of researchers worked with Teun Klapwijk, a superconductivity physicist, to build an expertise in organic crystal field-effect transistors. Hugh Hillhouse, a postdoc, set up a crystal-growth apparatus following the Bell Labs directions, and worked with a graduate student, Ruth de Boer, to deposit aluminum oxide and try to make field-effect measurements. Alberto Morpurgo, a research scientist, provided some hands-on supervision and a lot of furrowed determination.

Klapwijk had understood, from speaking to Bertram Batlogg in December 2000, that the Bell Labs work was being done on a run-of-the-mill sputtering apparatus that the group had found somewhere in the corner and used to lay down an aluminum oxide insulator. Researchers at Delft reckoned on doing the same, but they had the same problems as groups elsewhere. Ruth de Boer measured mobilities of charge in the Delft crystals that were around one hundred times lower than Schön had reported and found that their quality seemed to worsen over time. She was unable to deposit a layer of aluminum oxide without damaging the crystals' surface, and she and Morpurgo could not see how to avoid breakdown of the insulator at high electric fields either. But while acting as reviewer for a paper reporting on a superconducting-insulating-superconducting junction made from a field-effect transistor, Klapwijk noticed that the acknowledgment section thanked Ernst Bucher, the former supervisor of Jan Hendrik Schön, the first author of the paper, for lending the use of equipment. He gathered that this was a reference to Schön's having worked not at Bell Labs, but at the University of Konstanz in Germany.

At the time, Klapwijk assumed that Batlogg must have had some reason for having given him the wrong information about the location of the group's equipment. It seemed likely that there was a special trick to getting the experiment to work, that the Konstanz location was an important clue, and that Batlogg did not want to help potential competition by revealing it. It was time to do some detective work. Morpurgo followed up on the Konstanz lead by searching through old papers by Schön and Ernst Bucher. He came across a reference to Schön's diploma work on the deposition of aluminum oxide using a sputtering machine in 1994. The paper was in an obscure journal, so Morpurgo ordered a copy from the library. When it arrived, he couldn't find any information about laying down a layer of aluminum oxide that could withstand the high voltages that Schön claimed to apply to his field-effect transistors. Morpurgo later sent a disappointed note to Klapwijk: "They did not give any really useful information on the sample preparation part."

The group tried asking Schön whether he would like to come to Delft for a visit, but he said he was too busy. Hillhouse and de Boer got one chance to talk to him in August 2001, when they traveled to attend the annual "Enrico Fermi" Summer School for physicists on Lake Como in northern Italy. In 2001, the school was dedicated to the topic of organic materials, and Schön had been invited to showcase his results alongside researchers who had been recognized for years as pioneers in the study of organic materials, such as Norbert Karl. Hillhouse managed to sit and talk with Schön for an extended period overlooking the lake, he said. De Boer talked to Schön too. She found him modest, not at all bragging about his results, looking at others with curiosity, but not saying much. Later, Hillhouse sent an update to Morpurgo and Klapwijk. Encouragingly, Schön agreed that the mobility of charge in crystals degraded over time, as they had found in Delft, and that he had had problems recreating the mobilities reported in his original papers, as they had. Hillhouse said his own opinion was that maybe the variation could be explained by variability in the batches of starting material ordered from the standard chemical suppliers, Aldrich. (The same idea that was later mentioned by Batlogg to Joe Orenstein in Berkeley.) Although Schön had been reluctant to share much, Hillhouse had pressed him for some further details, and said he felt that they were getting closer.

But the new details did not lead anywhere either. At one advisory meeting to assess the progress of Ruth De Boer's PhD, Klapwijk recalled the student "practically in tears" as she presented her mobility measurements and described Jan Hendrik's as "better." De Boer said that in this period she was dejected and was thinking about giving up on her PhD topic, which would lose time. Klapwijk was trying to think differently. In Delft, there was an unusual requirement for the PhD that called for each student to propose five statements that could be about anything—science, culture, the arts, philosophy—and defend them in a debate with a committee of professors. It was supposed to teach a willingness to stand up to authority. The physics department also organized journal clubs, where graduate students were encouraged to present something that had recently been published and to criticize it. Klapwijk thought about questioning the assumption their research group had been working with: that Schön was the world leader and that they were supposed to catch up to what he had reported and get the same results as he had. Just because his data was published didn't necessarily make it "better." "I said, 'Ruth, maybe you're better than he is,'" said Klapwijk.

De Boer began to think about other ways to make her project into a success. She noticed that sometimes, when she picked up the delicate crystals, they stuck to her tweezers or to surfaces that she placed them on, held by the weak force of static electricity. That had given her an idea. There were at least two ways to make a transistor. Hendrik Schön claimed he put a crystal on the bottom, evaporated gold contacts, then sputtered on the gate insulator and added a gate electrode. But it was also possible to work upside down, putting down a gate electrode, then a gate insulator, then the crystal on top, and finally painting on some metal contacts. By putting the crystal on second to last, she would avoid having to process it by depositing the oxide. The samples' natural stickiness made her think that they would still adhere closely enough to the insulator to make a field-effect transistor. This turned out to be true, and she made her first field-effect measurements by gating the crystals from beneath. De Boer had rediscovered a common way of making transistors from delicate materials, the "flip-chip" technique. She drew balloons and fireworks in her lab notebook to mark the occasion.[7]

But because the mobility of charge in de Boer's crystals wasn't as high as Schön claimed, de Boer was never able to measure the sensational effects that Schön had reported. She never saw superconductivity or the fractional quantum Hall effect. She was able to measure ambipolarity, the conduction of current by either positive or negative charges, but only in a different single crystal from those Schön had claimed to study, copper- and iron- phthalocyanine.[8] Like many other groups, researchers in Delft continued work, figuring out which materials made the best contacts and insulators for use with plastic electronics. Morpurgo said that some of the research turned out to be easier with organic crystals than it would have been with plastic films, vindicating the rationale for having a research program such as the one conceived at Bell Labs. But, progress was still slower going than anything Schön had claimed. Looking back historically, it had taken years for researchers to refine silicon and develop the silicon transistor, and it was decades before the materials reached the point at which charge would flow at the high rates needed to support exotic quantum phenomena like the fractional quantum Hall effect. But because organic materials can be used to make a primitive field-effect transistor at a much lower purity than was possible with silicon, Schön's claim to have achieved so much seemed plausible. "There was a gap in our knowledge, and Hendrik drove a truck right through the gap," said Art Ramirez, whose group was trying to replicate Schön's work at Los Alamos National Laboratory in New Mexico.

At Rutgers University in New Jersey, Vitaly Podzorov and his PhD supervisor Mike Gershenson also worked for months trying to sputter aluminum oxide before Podzorov thought of changing techniques. Podzorov learned how to grow crystals from Christian Kloc's publications, but he had little back-and-forth with Schön. When he couldn't sputter successfully, he ran a short experiment, putting a crystal that he knew showed some electrical activity into a sputtering machine for a short time. When he took the crystal out, he found it was dead, and the aluminum oxide deposition hadn't even begun. From this, Podzorov realized that the mere presence of ions from the sputtering gun could damage the surface of an organic crystal beyond repair, and he dropped the idea of sputtering altogether. He switched to a different gate insulator, parylene, a plastic that could be shrink-wrapped in a thin layer, and the Rutgers group was the first after Schön to report high-quality field-effect measurements in organic crystals.[9]

But while some groups found a way forward by working around Schön's claims, others tempted into the field never got any useful results and gave up. At the University of California, San Diego, Bob

Dynes, a professor of physics and former Bell Labs department head who went on to become the president of the University of California, said that he also decided to get involved in the heady atmosphere of Batlogg's presentation at the March 2000 APS March meeting in Minneapolis, but didn't allocate 100 percent of anyone's time to what he regarded as a risky project, and later gave up. It was a similar story in Groningen, where researchers working with Thomas Palstra, a professor of solid state chemistry who had been a postdoc of Batlogg's at Bell Labs, tried to make field-effect transistors on thin films of pentacene without success. "We were struggling with all kinds of details. We felt like a bunch of amateurs being outcompeted by superman," Palstra said. But of the group did not assume that something was wrong with Schön's work, only that their lab couldn't build on it. "We were at a university. Bell Labs was a unique environment." Palstra assumed that Schön was very professional and had a lot of technical support. In other labs, however, including at the University of Florida in Gainesville, where researchers working with another former Bell Labs physicist, Art Hebard, tried to deposit aluminum oxide according to Schön's recipe, doubts began to emerge about the validity of Schön's work as a result of failures to replicate. Hebard said he accepted the idea that there was some kind of trick to the aluminum oxide deposition, but he could not understand how Schön was able to apply electric fields more than ten times those that his group could manage, "I started to have my doubts. I was jealous, and then I was suspicious," said Hebard. In Sendai, Japan, Katsumi Tanigaki of Tohoku University said that he discussed the problems with other researchers, heard a rumor that Schön could not replicate his work either, and decided that the problem was not in his own lab.

But at Stanford University in Palo Alto, California, David Goldhaber-Gordon, an assistant professor of physics, said that he would not have made the leap from his own failure to replicate a claim to the conclusion that the claim was wrong. Goldhaber-Gordon pointed out that postdocs and assistant or associate professors sometimes had to take a bigger chance by throwing themselves into new research areas. If they failed, they might assume that they needed to try harder to build up expertise. Goldhaber-Gordon was also in contact with Schön, meaning that, like the researchers in Minnesota, he was on the receiving end of a stream of encouraging, sensible-sounding advice. Goldhaber-Gordon's most experienced postdoc and a graduate student worked on organic crystals for nearly a year. When the fraud came to light, his student still had enough years left to switch to another project and graduate with a good dissertation, but Goldhaber-Gordon's postdoc, Silvia Lüscher (now Lüscher Folk) ended up having spent a year in her postdoc position without results, and, after leaving the lab, became less active in experimental physics. Although she had other reasons for this, the year of frustration had not helped. "You can't bring the time back," Goldhaber-Gordon said.

Researchers associated with Bell Labs also lost time following up, including Jochen Ulrich, a graduate student recruited by Horst Störmer to follow up on Schön's claims at Columbia University and Bell Labs. In late 2001, Ulrich's professor at the Technical University of Vienna received a call from Störmer, who asked whether he had a student who might be interested in coming to the States to explore the quantum Hall effect in organic crystals. Störmer had co-won his Nobel Prize for discovering the effect in gallium arsenide, so the chance to work with him on a different material sounded like a wonderful opportunity. In September 2001, Störmer and two other Columbia University professors were awarded more than $18 million by the U.S. National Science Foundation to build a Center for Electron Transport in Molecular Nanostructures at Columbia University in which, the proposal said, Störmer and others would, among many other projects, collaborate with Schön to build and characterize field-effect transistors. The proposal included $250,000 to buy a sputtering machine and funds for two junior postdocs to work jointly between Columbia and Bell Labs. Jochen Ulrich flew to the United States, visited Bell Labs, had an interview, and got a job. But to his disappointment, he never got to explore quantum effects in organic crystals and had to spend months on the mundane task of trying to remake Schön's field-effect transistors. Sick of the materials, he switched to a new topic but still left science at the end of his postdoc. "It was frustrating," he said, "I wouldn't go as far to say I left science because of this, but it changed the way I thought about science. I used to have a glorified view. Scientists search for the truth. That's not always the case." Part of the problem was that while researchers elsewhere gave up, those associated with Bell Labs were under the influence of information from Schön. When Podzorov came to Bell Labs for an interview in early 2002 to apply for a position as a postdoc to work with Schön, he warned Horst Störmer that he didn't think it was possible to sputter aluminum oxide, but found Störmer dismissive. Störmer told me by phone in 2005 that he and others had believed that Schön had found some way to mend the damage caused by sputtering, for example, by using hydrogen, which is what Schön had told researchers at the University of Minnesota too.

Claire Colin also spent around two years as a graduate student following up on Schön's work at the City of Paris School of Industrial Physics and Chemistry in Paris. Her professor, Michel Laguës, was part of a group that had provided samples of calcium copper oxide films to Schön, resulting in at least two publications. Colin was not part of the collaboration, but by September 2001 she was asked to try to reproduce the work in France, recreating both the films and the deposition of the aluminum oxide insulator. She succeeded in making the films, but the second part went nowhere. "We found that during deposition we were destroying the film, that the device was wrong in the conception," she said. "It was two years I could not use for my CV. It was not a good story to tell. It was my first real research experience." Fortunately, in the course of trying to follow up the work, Colin came into contact with Claude Pasquier of the Laboratoire de Physique des Solides of the University of Paris-Sud 11 in Orsay, where researchers had tried to sputter aluminum oxide without success. They understood the blank on her CV and offered her a position restarting her PhD from scratch on another topic. "I was lucky," she said.

Following the exposure of a major scientific fraud, senior scientists sometimes voice concern about the possibility of harm to the public image of their field. News reports and transparent investigations sometimes seem to be making the problem worse. But in practice, the scientists most seriously affected by Hendrik Schön's claims were more affected by the reality of fraud than the bad appearance that followed its exposure. Until the situation was clarified, they lost time and money, and many experienced unnecessary anxiety about their personal future or the future of their laboratories. Although they adopted very different strategies—some giving up quickly, others settling down to try harder—none reached firm conclusions about what was going on, and there were no charges of fraud made on the basis of irreproducibility alone.

Those that made progress following up on Schön's claims did so by being flexible. They rejected the authority of a famous institution and paid attention to their own experiences in the lab. They were open to anecdotes and rumors from others. Instead of trying to imitate Schön's approach, they developed their own. The self-correcting process happened, but it was haphazard and disorganized, with a lot of self-doubt along the way. In comparison to the mythical picture of researchers as an army of self-correctors marching in an organized way toward the truth, science is more of a guerrilla war.

THE FIRST INKLINGS OF FRAUD

In material that I saw, the earliest record of a suspicion of fraud was a note by the Nobel Prize–winning theoretical physicist Bob Laughlin. Laughlin wrote to a young materials science professor, Ian Fisher, on May, 18, 2001, a year before accusations against Schön were made in public. Laughlin and Fisher were both at Stanford University. Fisher had followed up on the Bell Labs claims by growing his first organic crystals and was eager to find someone who might be interested in making measurements on them. Laughlin wanted to warn that the deposition of the aluminum oxide insulator was likely to be tricky. Repeating rumors he had picked up at a recent conference, Laughlin told Fisher that the Bell Labs collaboration's success rate in creating working devices was supposedly 1/500, if that. "Many groups here, in Europe, and in Japan have tried unsuccessfully to reproduce this result. I am not the only person worried about fraud," he wrote.

While Laughlin felt that other people were as worried as he was, by most accounts he was the most vocal early critic of the Bell Labs claims. As early as August 2000, Laughlin interrupted Batlogg's presentation at the Kavli Institute in Santa Barbara by shouting "you should be ashamed of showing these plots."[10] In the presentation, Batlogg showed data that, like much of the preliminary data Schön circulated for feedback, covered only part of the range that could be experimentally explored, provoking Laughlin to complain that he couldn't properly judge the results. By the time of a summer workshop in 2001 in Aspen, Colorado, Laughlin's views had hardened to the point where he approached Takehiko Ishiguro, a Japanese theorist who had presented an explanation of some of the unusual features in Schön's superconductivity data, and, rather than commenting on his new theory, asked bluntly whether anyone in Japan had replicated the work. Laughlin did not see the job of a theoretical physicist as having to explain all experimental observations in terms of physical theory, but to scrutinize and discredit shaky findings and then work toward explaining the sound work that was left. His inquiry left Ishiguro with the impression that, in the United States, it was more urgent to figure out whether the Bell Labs results were real than to spend time trying to explain them. Ishiguro went on to become one of the first (I found) to report in writing that the Bell Labs claims were in doubt because not a single laboratory worldwide had been able to make a layer of aluminum oxide able to withstand the electric field that the group had reported.[11]

Laughlin's skepticism about Schön's data was due not only to the apparent lack of reproducibility, but was also based on an understanding of the science of electrical insulators. Even the best insulating material will fail when it is placed in an electric field strong enough to rip electrons from its atoms, setting them free to conduct current. This process, called avalanche breakdown, is a runaway chain reaction in which the ripping of an electron from one atom tends to trigger the ripping of several more. The greater the purity of an insulator, the more easily breakdown can occur because impurities and imperfections in a material tend to create blockages and jam the avalanching process. On this basis, Laughlin would have expected a gate insulator as good as the one Bell Labs had developed to be a dirty and amorphous material.

But at the time Laughlin first heard Bertram Batlogg speaking about Schön's data, Batlogg was emphasizing the high quality and purity of Christian Kloc's crystals, as well as that of the interface between the crystal and the insulator. "He said that the interface was perfect," Laughlin recalled, and Batlogg's taped address at the Kavli Institute in California conveys a similar impression.[12] So, Laughlin was skeptical not only because the group's electric fields were greater than those withstood by the best insulating aluminum oxide, but also because Batlogg was not presenting the physical picture that Laughlin would have expected to hear from a research group that had truly developed a groundbreaking insulator. Thereafter, Laughlin told me, he focused on the breakdown in aluminum oxide as if it were the primary breakthrough they were claiming, aiming, he told me, to discredit the group by deliberately highlighting what he saw as the weakest part of their scientific case.[13] Laughlin said that at an early stage, he asked Batlogg to show additional data that could cast light on the properties of this unique insulating layer: when he didn't see that, he assumed that the work was fraudulent. By this, he didn't necessarily mean he knew Schön's data were fake, but that, at a minimum, Bell Labs was either failing to do and present the research necessary to confirm or falsify its own claims, or failing to reveal the tricks being used, a position that he regarded as entirely unscientific.

Although this way of thinking had propelled Laughlin to the right answer, his views on fraud did not gather momentum in the scientific community straight away. There were several reasons for this. One was that Laughlin was known to apply very strong language in cases where he detected a hint of wishful behavior. Stephen Forrest, who was himself a fierce skeptic regarding the validity of the Bell Labs laser data, said bluntly of the possibility that Laughlin had realized it was fraud early, "Laughlin doesn't count," because, Forrest said, Laughlin was known for being too skeptical. In addition, other researchers in organic electronics were likely to have regarded Laughlin as peripheral because he was not known for working in the area,

Forrest said. Third, most scientists found it inappropriate to talk about fraud simply because a result seemed generally unconvincing. Forrest said that in his own laboratory, the possibility of fraud sometimes seemed to hang in the air between himself and his student Marc Baldo, who was reanalyzing Schön's data, but that he took a principled view that it was not possible to second-guess Schön's integrity on the evidence they had, and encouraged his students to do the same. If Baldo questioned Schön's claims too strongly, Forrest said, "we can't think like that," so that the discussion never turned to the possibility of fraud.

In practice, however, Bob Laughlin's position was more similar to that of Forrest than his remarks might have suggested. Although Laughlin had mentioned fraud in a private note, he did not make a formal allegation of misconduct to managers at Bell Labs, and he did not try to search for and publicize specific evidence for fraud in Schön's work. Ian Fisher, who received Laughlin's email, said that the mention of possible fraud might have raised a flag, but that he did not remember it deterring his crystal growth efforts.

Although Laughlin's accusation did not stick, his position that the Schön collaboration needed to clarify the aluminum oxide deposition did, and it was also independently voiced by others. Another early Schön skeptic was Ivan Schuller, an experimentalist at the University of California, San Diego, who had worked with aluminum oxide and knew from his experience that it could not withstand the high fields Schön reported. Schuller was a vocal skeptic at conferences and workshops, and one physicist, Laura Greene of the University of Illinois at Urbana-Champaign, recalled that after she spoke with Schuller, she felt very concerned. In November 2001, she raised concerns about the explicability of Schön's data to Batlogg, who was a friend of hers. George Sawatzky, a physicist at the University of British Columbia in Canada who had long been interested in organic crystals, was also asking a number of questions about the way in which Schön's devices were made. Sawatzky was supportive of the Bell Labs work, but he had developed concerns about its accuracy after Batlogg was unable to respond to his question about the orientation of organic crystals in some of the reported field-effect devices, apparently because Schön had admitted that he had not kept a record of the information. Sawatzky regarded the deposition of the insulator as key to understanding the devices, in part because Schön's incredibly smooth curves showing the resistance of a superconducting carbon-60 field-effect transistor showed no sign of subtle features that Sawatzky expected to be present. Not realizing that this was because Schön lacked an awareness of the physics underlying those features, and so had fabricated nice smooth doubt-dispelling data that did not include them,[14] Sawatzky and a colleague began running calculations to see whether the presence of the amorphous layer of aluminum oxide could have caused a change in the behavior of carbon-60 and thus explain the unexpected smoothness.[15] Sawatzky also continued to question Batlogg about how the devices were made.

Facing such inquiries, Schön and his collaborators gradually began to describe his sputtering efforts in a different way. By 2001, they were likely to respond to questions by saying that the aluminum oxide was something they needed to understand better, and that Schön could not always reliably produce it. In turn, this position contributed to the rumors that other researchers were hearing. The idea also spread that Schön had been working on aluminum oxide deposition for several years and was expert at it, and that he was using a machine in Konstanz that had been used to sputter other materials in the past, and so contained contaminants that would result in an amorphous, and even dirty, deposition, not an ultraclean one. Batlogg and Kloc even reasoned that this might help to explain why others were having difficulty replicating Schön's work; they were not so expert, or their equipment lacked the unique, unknown history of the Konstanz sputtering machine.[16] So although the lack of reproducibility did not result in public allegations of fraud, it did put pressure on Schön's collaborators to begin to confront the extent to which their apparently ordinary coworker was turning out to be such an extraordinary success.

8

PLASTIC FANTASTIC

Toward the end of his time as a postdoc, Hendrik Schön approached Christian Kloc for advice. Schön explained that Bertram Batlogg was leaving Bell Labs for Zurich, and had asked Schön whether he wanted to come too. As Schön had been working hard toward a permanent job at Bell Labs, he did not know what to say.

Kloc told Schön not to go. Schön had Kloc and the materials that he needed readily available at Bell Labs. Setting things up in Zurich would take a while, Kloc explained. Kloc also pointed out that because Batlogg was very well-known, and had been giving many public talks about their collaboration's research, it was likely that outside researchers were giving Batlogg much of the credit for Schön's work. If Schön moved with Batlogg to Zurich, he might have more trouble being fully recognized as an independent researcher, Kloc warned. These were fairly typical considerations to mention to a junior scientist, but Kloc saw the issue as complex for Schön, because of the apparent importance of his work. "I said, 'Hendrik, you're the star,'" Kloc told me.

Following the advice, Schön stayed at Bell Labs, and in December 2000 got his permanent job as a member of technical staff. To take the job, he needed a new work visa, so he flew to Germany, where he worked for five months in Ernst Bucher's laboratory before, he said, his visa arrived. This was only one of multiple trips that Schön had been making to Konstanz, but the visa became a convenient excuse both for him and, after his disgrace, even for managers, for his having worked off-site.[1] This break was longer than most of the others, and Schön took advantage of the opportunity to strike out as a "star." Out of the tourist season, Konstanz was more low key than ever, with the lakeside promenades mostly empty and students able to ride unobstructed through the old town on their bikes. Schön worked on the computers in Ernst Bucher's lab, regularly sending manuscripts and data back to managers at Bell Labs.

By this point Schön was thirty years old. He spent time in Konstanz with a group of Ernst Bucher's graduate students, including Ortwin Schenker, who had been given Schön's responsibility for the lab's photoluminescence apparatus when Schön left. The two shared photoluminescence data that appeared both in Schön's thesis in 1997 and in Schenker's in 2002, and when Schenker arranged to take a year out as a visiting researcher in South Africa with Vivian Alberts, a former student of Bucher's who had set up a solar energy technology lab at Rand Afrikaans University (later part of the University of Johannesberg) Schön joined him for a research visit. But to Schön, Schenker was more than a friend; he was also a fellow CGS-er. Schenker and others in Bucher's lab spent many weeks following up on Schön's claim of n-type current—current carried by electrons—in copper gallium selenide, but did not have much success.

A web site registered during Schön's extended Konstanz stay gave some clues about his possible range of activity. The site was located at www.ratskrone.de. Ratskrone is a cheap German beer. The main image on the site showed a short-haired, muscular young woman wearing a blue satin dress, white pearls, and a sash on which was written in gold handwriting, "Miss Ratskrone 2000." All around were photos of empty Ratskrone bottles, and a sequence of shots showing Miss Ratskrone flushed in her rouge and pouting as she opened beer bottles at a dark, depressing bar. At the bottom, a fake ticker recorded that the site had had 8822 visitors.[2]

Against the background of a preexisting collaboration, it would not have been hard for Schön to involve Schenker in his Bell Labs research projects. During Schenker's time as a student, Schön had been making regular, short visits back to Konstanz, after each of which he returned to Bell Labs and claimed to have made experimental progress in sputtering aluminum oxide. The visits had not been an issue at Bell Labs until outside researchers started to ask about sputtering. But the most immediate consequence of their inquiries was not that Schön scaled back his research claims, but that he began to think of himself as a Konstanz-based sputtering expert. He could rely in part on a reputation for experimental skill that was already being built up by senior colleagues. For example, Horst Störmer had advised colleagues in 2001 that he regarded Schön's work as a kind of "wizardry," because, he said, Schön was able to cleave organic crystals into fragments that were as small as half a millimeter across, and turn those into field-effect transistors using only his hands, a feat that Störmer claimed he had tried himself and could not replicate. Störmer added that although he'd never seen Schön do this, he didn't blame the younger man for not wanting to be watched and that the devices worked. Bertram Batlogg similarly advised Bill Brinkman, who was vice president for research at Bell Labs until September 2001, that Schön could do experiments no one else could because he had "magical hands," Brinkman told me. Schön steered clear of making such extreme claims about himself, but he did not go out of his way to contradict others who did and even made an effort to support them. In 2001, he added information about his 1993 diploma project on the sputtering of solar cells into a short scientific bio,[3] even though this project had never been considered an important part of his research experience before. He began to call his aluminum oxide layer "amorphous,"[4] to put across the idea that he had worked carefully to produce the chaotic disordered structure that experts figured such a material must have in order to withstand the high electric fields he claimed that he had applied. He told George Sawatzky, a physicist at the University of British Columbia, that he had spent his PhD days sputtering, even though his PhD had focused mostly on photoluminescence. He mentioned to Doug Natelson of Rice University in Houston that setting up a sputtering apparatus from scratch involved careful optimization and took a long time. Even disagreements about equipment became opportunities. When Art Ramirez, whose group was trying to replicate the work at Los Alamos National Lab in New Mexico, asked how far Schön's sample was from his sputtering gun,

Schön said that it was about eighteen inches. Ramirez responded that the sputtering machine at Los Alamos was not large enough to place a sample eighteen inches away from the gun, and that the Los Alamos group was sputtering from a distance of about six inches. Schön said that yes, the sputtering machine in Konstanz did have an unusually large space inside it, one advantage of which was that he could sputter more gently than others could. Following the discussion, Schön wrote the figure of only four centimeters in his draft manuscript about sputtering,[5] although without telling Ramirez that he had made a change.

While Schön was building up his reputation for aluminum oxide sputtering, he also arranged for Ortwin Schenker to provide some technical assistance. Because most of Schön's visits to Konstanz had been brief, the number of claims Schön could make about the results he had obtained there was limited. But if Schenker could sputter samples for Schön, and then give them to Schön to measure, Schön could work faster than if he had to claim to be doing all the sputtering himself. At the same time, it became increasingly clear that there was no reason why Schön had to confine himself to working only with organic crystals. The provision of samples by Christian Kloc had been a frustrating limitation on the pace at which Schön could make new claims.[6] Given that Schön was a sputtering expert, he could arrange for other materials to be sputtered too.

Schön's first high-profile publication with Schenker reported on an attempt to turn the common plastic polythiophene into a superconductor. The project followed the award of the 2000 Nobel Prize for Chemistry for the discovery of plastics that could conduct electricity, so Schön was sure that a positive result would seem topical. At Murray Hill, he was getting to know Zhenan Bao, a chemist who was gradually moving into the field of plastic electronics. He set up a collaboration that spanned two continents as he took samples sputtered by Schenker in Konstanz to New Jersey for Bao to deposit a layer of polythiophene. Finally, he took the samples back, added metal contacts, and claimed to apply an electric field through his aluminum oxide layer, drawing charges onto the surface of the plastic until the material went superconducting. One advantage of this method, Schön explained to one outside researcher, was that he could avoid reporting sputtering on top of the delicate plastic, a step that he knew was causing trouble for other scientists. For several months this arrangement left unclear to what extent Schön's success was due to the "magical hands" and to what extent it was due to the sputtering equipment in Konstanz, which Schenker could apparently operate too.

By the time the superconducting polythiophene appeared in *Nature* in March 2001, Schön was in Konstanz in the middle of his five-month stay. After the online magazine *Physics Web* coined the term "Plastic Fantastic" for the breakthrough, Schön decided to borrow the phrase for the title of a talk that he delivered at the New Jersey branch of the American Chemical Society.[7] It was a more excitable title than he would normally use, but Schön was beginning to learn that even a scientist who is personally very nice and modest must make sure to show some strong enthusiasm for the science he reports. His result was a personal tour de force, as it showed that he had finally succeeded in reporting developments in industrially relevant plastics. When Lucent's vice president for corporate strategy, William O'Shea, wanted to convince a computer industry reporter that Lucent had an edge over competing telecommunications companies, he pointed to the research on plastic superconductors as an example of the way that the company's support of fundamental science at Bell Labs was leading the company to "future products or projects."[8] Again, Schön was not interested in the money; his paychecks were piling up unopened on his desk at Murray Hill all the time that he was in Konstanz. When Kloc called and offered to take them to the bank, Schön agreed. When Kloc said one paycheck was missing, Schön said not to worry.

Money was not important to Schön, but fulfilling the expectation that he might strike out alone was. In 2001, Schön learned that his papers in *Nature* and *Science* had come to the attention of Catherine Deville Cavellin, a professor at the University of Paris 12 Val de Marne. She was working with thin films of calcium copper oxides that were expected to become superconducting at unusually high temperatures, although researchers working with her had not had much luck observing the phenomenon in practice. Given that Schön had achieved a first by inducing superconductivity in organic crystals, Deville Cavellin wondered if he might be able to do the same with her films. Superconductivity in cuprates often depends on the introduction of impurities that change the amount of charge available in the material but can disrupt the internal structure of the materials; Schön's technique of adding charge using an electric field applied through a layer of sputtering seemed a way around this.

For Schön, the approach from the French group provided a way to strike out independently from Kloc at Bell Labs by receiving some samples from elsewhere. He agreed to try out the French experiment and soon received an old calcium copper oxide sample made by a graduate student working with Deville Cavellin at his address in Murray Hill. Within a few weeks, he said that he had seen superconductivity at 75 degrees above absolute zero, higher than had been seen for other similar materials. This was a big result for the French, who had been working toward superconductivity in such materials for nearly a decade.

Schön knew, of course, that the French had been reading his organics work, and the data he used for his positive result in the copper oxide materials was similar[9] to that which he used for a different line of work with carbon-60, or buckyballs, soccer ball–shaped molecules of sixty carbon atoms. In May 2001, he produced data showing superconductivity at 117 degrees above absolute zero, a result better than anything expected by a colleague, Martin Jansen at the Max Planck Institute in Stuttgart, who had given him the idea for the experiment.[10] He consolidated the claim by giving a printout of the data to one student in Konstanz and inviting several people out for a drink to celebrate, he wrote in an email sent later to a colleague. Although making a claim in real time might seem risky, Schön wrote, again in an email circulated after his disgrace, that the Konstanz students were anyway "no good for being witnesses," because they were working on solar energy technology. Because their PhD thesis topics were unrelated to organic electronics, he said they felt they lacked the expertise to judge his claims.

He promptly sent manuscripts to *Nature* and *Science* about his results on the copper oxide and carbon-60 samples. For the first paper, he gave a corresponding email address, the one that outside scientists would have to use to write to him with questions, of "j.hendrix@ratskrone.de." This was apparently a real email address, but it also contained an apparent pun on his name, "Jan Hendrik," and that of the American guitarist "Jimi Hendrix," transported into the context of the drink preferred by Miss Ratskrone. The strange mix of cultural references may not have been appreciated by editors at *Science,* but it did not matter, as the paper appeared back-to-back with Schön's latest result on superconductivity in carbon-60, where he was listed as the company man, "hendrik@lucent.com."[11] Still, some people picked up a sense that Schön was struggling to resolve his new identity as a Bell Labs star with his old one as a regular German student. If it is true that, as Charles Babbage suggested in 1830,[12] there is a difference between a forgery and a hoax—a hoaxster wants his scam to be uncovered, to prove a point to those he has hoaxed, while a forger wants to remain secret forever—Schön was teetering on the edge between a bold forger and a shy hoaxster. When few people noticed the email address, he mentioned it himself, but said that he had gotten into trouble for it. One Konstanz colleague half-laughed as he told me how he had wondered whether the Hendrik they knew was becoming an extrovert. For sure, Schön's attitude to publication had developed in this direction: while in late 1999 he had expressed hesitation to a coauthor over a submission to *Science,* by 2001 it was he who eagerly suggested to the French that they try their luck at the journal.

Striking out alone wasn't easy, though. In October 2001, the city of Braunschweig in Lower Saxony, the German state in which Schön was born, selected the Schön–Kloc–Batlogg collaboration to share the prestigious Braunschweig Award, a DM 100,000 prize for science. Schön's mother and stepfather traveled proudly from Austria to Lower Saxony for the ceremony. Yet both the award ceremony and the press coverage focused on Batlogg, so that Schön had to listen as Batlogg made an acceptance speech and, according to the way that comments were excerpted for a brochure placed on Lucent Technologies' German web site, explained that the research program had succeeded because "we haven't asked for permissions or stuck to plans. Our only limitation is our imagination."[13] When Schön was recommended for the 2001 William L. McMillan Award, which is given every year by the University of Illinois physics department to an up-and-coming young physicist, but lost out because, Batlogg was informally told, the work for which the young German was known had not been reproduced.

Schön behaved as if he did not care, telling Ernst Bucher that it was other scientists' problem if they could not replicate his work. In Konstanz, he operated in a supervision vacuum. Because Bucher was not a coauthor on papers involving sputtering he was not supervising Schön's work on the technique. Schön's managers and collaborators from Bell Labs did not visit to watch or supervise, and when Bertram Batlogg came from Zurich, he came only, as Bucher put it gruffly "to pick up data." If there was the potential for territorial tension there, Schön leveraged it. During Bertram Batlogg's visit, Schön shrank back into his role as Bucher's former student, showing Batlogg the photoluminescence apparatus for which he had once been responsible and keeping well away from the sputtering machine.

But Schön could not keep Batlogg away from sputtering forever. By the end of the summer, questions from outside groups, and Batlogg's desire to set up his own sputtering apparatus in Zurich, caused Batlogg to ask more questions of Schön than he had done previously. Batlogg has insisted to me that around this time he also began to ask Schön to explain the recipes he was using in greater detail. By October, Schön had produced an early draft of his technical manuscript on sputtering and documented discussions with Kloc and Batlogg in it. Batlogg's lab in Zurich was also joined by visiting researchers from Japan who wanted to understand the Bell Labs claims. Jun Takeya, a research scientist at the Central Research Institute of Electric Power Industry (CRIEPI) in Tokyo, and Tatsuo Hagesawa, from the National Institute of Advanced Industrial Science and Technology in Tsukuba, came to Zurich to try to learn how to make field-effect transistors with organic crystals. Batlogg had a graduate student, Claudia Goldmann, who was doing her thesis on this topic, and a postdoc, Christoph Bergemann, who had joined the lab because he was interested in following up on the work with

Schön too. Like other lab heads who wanted to build on Schön's claims, Batlogg purchased the equipment Schön claimed to have used, a Keithley electrometer for making measurements and a Leybold sputtering machine for laying down the aluminum oxide. Batlogg had to buy the machine second-hand because the model was no longer available new. But Zurich is only an hour from Konstanz by train, and Batlogg's junior researchers went to check things out. As Ernst Bucher's students had apparently not yet realized the extent to which the international scientific community had come to think of the Konstanz-based equipment as unusual, the sudden appearance of researchers who were very curious about such commonplace equipment did raise some eyebrows, but the visitors were quite welcome to come in, sputter, and run experiments in Bucher's lab. Schön helped out at some of the sputtering demonstrations in Konstanz. When the tests did not result in functional devices, he told Batlogg that it appeared Ortwin Schenker was having trouble with the sputtering machine.

These efforts by Schön's collaborators and Batlogg's junior researchers to try to get Schön to replicate his work contributed to some of the rumors in the scientific community that had influenced Bob Laughlin, as well as researchers in Japan, to be suspicious that the work was invalid. Inquiries from outside researchers also had an impact in forcing Schön to back down from his earlier claims. In May, Doug Natelson, a professor of physics at Rice University in Houston, Texas, who had been a postdoc at Bell Labs, wrote to Schön and Batlogg and asked whether they would be prepared to sputter some aluminum oxide onto blank pads of silicon, which Natelson's group could process into field-effect devices. In response, Schön told Natelson that his sputtering system had not been operational since around February 2001. Batlogg's response was more vague, but also conveyed that the set-up was out of commission. "We are in the process of changing our set-up and are out of business for a while. As you understand, this is obviously an important aspect of the work and we would like to push it further," wrote Batlogg.

As researchers in Zurich had trouble following up Schön's work, Batlogg said he became uneasy and asked Schön why he was having so much more luck than others, and whether there were materials that it had not been possible to make into field-effect transistors using sputtering. At this, Schön, who had previously focused on circulating news of positive data and good results, willingly sent Batlogg a list of failed experiments. Batlogg was reassured that even in Schön's hands, the technique was sometimes a little tricky.

As concerns about sputtering mounted, Schön approached his former postdoc supervisor with a new line of distraction. At Bell Labs, Schön had been talking with Chandra Varma, a senior theoretical physicist and MTS who was very interested in the data on the cuprate samples from Paris. As Varma explained it to me, research on cuprates has been limited by the need to synthesize a variety of different structures, each containing different combinations of copper and oxygen atoms, and each of which becomes superconducting at different transition temperatures. But Schön was claiming that he could use the field effect to change the density of charge in one sample, sweeping it through a wide range of different properties and transition temperatures, as if he had hundreds of samples at his disposal. Following discussions with Varma, Schön came up with detailed data showing the transitions between superconductiving, conducting, and insulating behavior. as the surface of the cuprate samples was tuned by the field effect. Varma was more than impressed. "I was thrilled. I had believed such a thing would exist. It was so systematic," said Varma.

Schön could rely on Batlogg sharing this sensibility. In 2000, Batlogg and Varma had written an article for the magazine *Physics World* about unsolved problems in cuprates research, which Schön read and took as partial inspiration for another manuscript.[14] From Zurich, Batlogg had been uninvolved in Schön's work on cuprates, but at the end of the summer of 2001, he came to Murray Hill for a visit and became involved in discussions with Varma. "We were chattering like the old cuprate soldiers," said Batlogg, referring back to the days of high-Tc, when cuprates first burst onto the international scientific stage. Schön followed up on this vulnerability with a request to Batlogg to take the position of corresponding author on his next manuscript about cuprates because, he pleaded, he had so much work arising from questions on earlier papers, an apparent reference to sputtering. This might have been a red flag: given that Schön had told outside researchers that his sputtering apparatus had not worked since February, it is hard to understand how he could have done any of the work on cuprates, samples of which had been sent to him after March, Deville Cavellin recalled. But while Batlogg had known of the sputtering problems in May—because he had told Natelson about them—he had not been approached about the cuprates data until late summer. Schön's distraction tactic meant that Batlogg oscillated back and forth between enthusiasm for the new data and concern about getting the sputtering to work reliably. In September, Batlogg presented some of Schön's data on carbon-60 at a Gordon conference in Oxford. There he was approached by John Wilson of Bristol University, who asked about the possibility of using the field effect with cuprates. Batlogg had not shown cuprates data at the meeting and apparently replied that the method might not work with the materials. But in December 2001, Batlogg showed cuprates data acquired by Schön at a symposium organized by the Nobel Foundation in Sweden, where many of those present were Nobel Laureates. Philip Anderson, a theoretical physicist at Princeton who had won a Nobel Prize in Physics for his work on disordered materials, said that he and others took the incredible sweep of Schön's data seriously only because of Batlogg's stature in the field of high-temperature superconductivity.

Schön's attempt to make a name for himself independently of Batlogg had failed, but he was having success in distracting colleagues' attention from the sputtering problems. He was very much enjoying traveling back and forth, and he happily mentioned his busy schedule to anyone who tried to engage too closely with him at one site or another. When Deville Cavellin tried to arrange a meeting in Europe, Schön said he was too busy, which caused Deville Cavellin to talk about coming to Murray Hill. But colleagues at Murray Hill did not have any more success accosting Schön, as he slipped away effortlessly after seminars and meetings. Even at parties, he preferred not to talk about work, which students at nearby Princeton University found very unusual for someone who was so successful. Even when he had a drink, he did not loosen up and start talking about work, in the way that someone who was only shy might have done. At one house party in New Jersey in 2001, Schön could not bring himself to mix, but stood on a porch all on his own, looking down on Bell Labs friends and colleagues who were playing in the garden with an electric car. When the wife of one of his collaborators noticed him there and commented that it looked as if the success had gone to Hendrik Schön's head, she was told that if only she knew Schön from coming into the lab every day, she would know that he was very modest.

9

THE NANOTECHNOLOGY DEPARTMENT

While Hendrik Schön was in Konstanz in early 2001, much of the data he sent to his managers at Murray Hill demonstrated his progress in the field of nanotechnology. "Nanotechnology" is a broad term that can be applied to any research that has involved measuring matter on scales as small a nanometer, a billionth of a meter. Pundits have long enjoyed imagining what would be possible if scientists could manufacture matter at the nanoscale—from the 1967 movie *Fantastic Voyage,* in which a submarine is shrunk down and placed in the bloodstream of a patient, to Eric Drexler's suggestion in his 1986 book *Engines of Creation*[1] that self-replicating nanoparticles might turn our planet into a mass of "gray goo." A more practical goal for nanotechnology researchers today is that of making electronic components, like wires, transistors, and computer memory, out of individual molecules. This subfield of nanotechnology, called molecular electronics, came to the fore in the late 1990s, just as Hendrik Schön was casting about for new research areas to dive into.

Molecular electronics gained in prominence because of limitations on the future potential of silicon electronics. Silicon chips are made in a process called optical lithography, which uses light beams. Just as a blunt crayon cannot be used to make a fine line drawing, so chip manufacturers cannot etch out transistor channels finer than several hundred nanometers, the wavelength of light. In addition, the gate insulators of transistors cannot be made thinner than a few atoms without current leaking through them. Experts therefore expect a slowdown in the growth of the computer industry early in the twenty-first century. Moore's Law, the famous observation that the number of transistors on each computer chip has doubled every couple of years as the computer industry has developed, is eventually expected to break down.

Molecular electronics was a way to go beyond this, building computers from the bottom-up. The history of the field is often traced back to 1974, when Mark Ratner of New York University and Arieh Aviram at IBM Thomas J. Watson Research Center in Yorktown Heights, New York, showed that if an organic molecule could be precisely controlled by an external electric field, it could act as a transistor tens of thousands of times smaller than today's.[2] But it took until 1997 for researchers to make practical progress toward the dream of connecting molecules together to form nanoscale circuits. That year, Mark Reed at Yale University and his colleagues claimed to have built a molecular junction in which molecules of the organic molecule benzene–1–4-dithiol were used to transport current between two gold electrodes.[3] In 1999, Reed claimed to have measured negative differential resistance, a molecular switching effect in which an increase in voltage caused a decrease in conductivity, in another organic molecule.[4] The same year, James Heath at the University of California, Los Angeles, and Stanley Williams at Hewlett-Packard Labs in Palo Alto, California, reported that they had made logic gates, the parts of electronic circuitry that process information, using rotaxanes, dumbbell-shaped organic molecules.[5]

The last two claims, both published by *Science,* proved controversial, in part because interpreting the data from nanoscale devices involves making assumptions about what has been built, leaving plenty of room for uncertainty. But the claims nevertheless prompted tremendous media interest.[6] And as nanotechnology emerged as a high-profile field of research, it also got onto the radar of researchers at Bell Labs. David Abusch-Magder, an MTS who began to move into the area even before Hendrik Schön, explained why. "Molecular electronics was an interesting perspective, thinking about electronics from the bottom up, measuring a small aggregation or one molecule. We also had a group of exceptionally good chemists. If there was the right place and right time, all the structural ingredients were there for progress to happen."

At the same time, physicists at Bell Labs came under more pressure than ever to demonstrate their lead in high-profile areas of science. In 2000, Bell Labs hired a new manager, Jeff Jaffe, who in 2001 replaced Bill Brinkman as vice president for research. Cherry Murray, who reported to Brinkman, and later to Jaffe, was asked to define and set goals for research more tightly than had been done before. "We were in economic freefall. We weren't exactly sure what amount of money we had but we knew it was going down. I set up a research strategy process every year. What kind of science should we be doing, what kind of research should we be doing," Murray said.[7] Pressed by an editor for *Physics World* in 2001, Murray admitted that researchers in her division were feeling the pressure.[8] Nanotechnology was not only a high-profile area, it was one in which Bell Labs could apply for funding independent of Lucent revenues because U.S. government agencies were interested in the area too. In 2001, Bell Labs was named as an external center of collaboration in proposals for grants by Columbia University in New York City, which was pitching for a $10.8 million Center for Electronic Transport in Molecular Nanostructures to the U.S. National Science Foundation, and by Brookhaven National Lab, which was applying for a U.S. Department of Energy $85 million Center for Functional Nanomaterials.[9]

Hendrik Schön was listed as a participant on both of these proposals, and from 2001 onwards frequently expressed his eagerness to become more involved in nanotechnology. His interest was first provoked some time in the latter half of 2000, just as the profile of the field was becoming impossible for Bell Labs researchers to ignore. In June 2000, "the birth of molecular electronics" and a popular article written by Mark Reed and his Rice University collaborator James Tour made the cover of *Scientific American.*[10] In November 2000, a review of molecular electronics by Christian Joachim at the French National

Center for Scientific Research in Toulouse, Aviram and Jim Gimzewski at IBM's Zurich Research Laboratory made the cover of *Nature.*[11] Schön's paper claiming to have seen superconductivity in a carbon-60 field-effect transistor appeared in the same issue and was given a mention on the cover, but clearly seemed less exciting to editors than the newer area.

To dive into nanotechnology, Schön had to leave his old collaborators, Christian Kloc and Bertram Batlogg, behind. Kloc's organic crystals were millions of times larger than the nanoscale, and Batlogg's view, held in common with other establishment physicists, was that nanotechnology was a big name, but not a big deal.[12]

But Schön was beginning to orientate himself to a different group of colleagues. After January 2001, he was listed on company documents as an MTS in the department of John A. Rogers, a busy and ambitious manager who had always been involved in the business side of science. After getting a PhD in physical chemistry from the Massachusetts Institute of Technology in 1995, Rogers, his supervisor, and another graduate student cofounded a company that they later sold to the electronics giant Philips. Rogers spent two years working in the Harvard laboratory of the chemist George Whitesides and three years leading an effort in plastic electronics as an MTS at Bell Labs before being promoted into the ranks of management in August 2000. Commenting to me in 2005, Rogers suggested that Schön had joined his department around two months later and that he had no information about what Schön had been doing prior to the summer of 2001. Chronologically, this is hard to reconcile with Schön's having been in Rogers's department from January 2001 onwards—but what came across was that Rogers had felt himself to be an inexperienced manager who was not on top of the replicability problems arising from Schön's work as a postdoc straight away. From January until May 2001, Schön had been working for Lucent from Konstanz, where he claimed to be waiting for a visa, rather than at Murray Hill. It was also common for post-doctoral researchers who were hired as MTSs to change departments and fields of research. Schön's move into Rogers's department therefore provided Schön with an excuse to dive into nanotechnology, effectively abandoning obligations associated with his publications on organic crystals. He seemed a big catch for Rogers's small department, which until the summer of 2001 was called "The Department of Condensed Matter Physics Research."

Along with talking to Rogers, who became involved in co-conceiving Schön's first nanoscale devices, Schön also formed a new collaboration with Zhenan Bao, the chemist who had worked with him on the "Plastic Fantastic." Bell Labs researchers had always been encouraged to set up collaborations that bridged the divide between conventional disciplines like physics and chemistry and Schön and Bao's collaboration was soon highlighted by managers as a leading example of this. Bao declined to talk with me about her involvement with Schön, but her department director, Elsa Reichmanis, explained that Bao's background had been in organic chemistry before she moved into plastic and molecular electronics. "Since that time she's developed some expertise, and with any kind of materials engineering some basic knowledge of characterization of devices is beneficial," said Reichmanis. But Bao lacked this basic knowledge when she got started. When moving into a new area, an MTS did not receive formal training because there had traditionally been a wide range of expertise on tap at Bell Labs. "You read the literature, talk to colleagues, and you do it," said Reichmanis. This meant that even though neither Schön nor Bao had much of a background in molecular electronics, their managers were delighted for the two young MTSs to set up a new research effort in the area. Rogers later told a reporter that the collaboration happened "spontaneously," springing out of the type of lunchtime discussions that were part of everyday life at Bell Labs.[13] But while managers might have seen things that way, Schön was far from spontaneous, always carefully pitching himself to fit in with his understanding of managers' strategy for research, while Bao was motivated at least in part by her perception of nanotechnology as a high-profile area. Peter Ho, a visiting scientist from Cambridge University said that he was skeptical about some of the preliminary data that Schön had produced, and he turned down an invitation from Zhenan Bao to join the collaboration.

Ho said Bao had tried to make the argument that it would benefit him to get involved, because molecular electronics was such a hot area.

"THE DATA CAME OVER THE ATLANTIC OCEAN"

Schön made it look easy. He placed an order for some new organic molecules, apparently following Bao's advice for which would be good to use, then flew to Konstanz and started sending back data. It wasn't long before his first results came to the attention of Federico Capasso, who as vice president of the Physical Research Lab was the manager between John Rogers, Schön's director, and Cherry Murray, who was responsible for setting research strategy. Early in 2001, two MTSs in the Physical Research Lab, Rafi Kleiman and Bob Willett, were standing in a corridor in Research Building One when Capasso came by excitedly wielding a couple of viewgraphs. It was data from Hendrik Schön apparently showing the operation of a nano-sized transistor made in Germany. "That was how the first information came to Murray Hill. It came not to colleagues, but to the director. It came over the Atlantic Ocean," said Kleiman. Other MTSs remember rumors of the breakthrough spreading in the cafeteria at lunch. Kleiman said he found what was happening unusual. "At Bell Labs, I'd the pleasure, a number of times, of being called into a colleague's lab" to see an experiment, said Kleiman. Other staff or nearby managers would scrutinize the experiment, and news of good results would then filter up to senior managers. Instead, with Schön off-site, his data appeared to be flying up the management chain, bypassing some of the normal collegial and scientific channels. This impression was not helped by the perception of Capasso as a manager who was always looking ahead to the next publication or press release. One former MTS told me that in this period he stopped telling managers about his research ideas until they were ready to be submitted for publication because he did not want to feel rushed. And John Rogers at a later point also asked Capasso to be careful in handling a draft of a Schön paper that Schön had not yet circulated widely.

Schön's first nanotechnology results were handled less cautiously. Within a couple of months of data coming back from Konstanz, Capasso had a slideshow ready. The slides, originally from Schön, emphasized the cross-disciplinary collaboration of Bao and Schön and declared the existence of a forthcoming development that promised to "revolutionize" the fields of organic and molecular electronics. According to these slides, molecular electronics could be compared to conventional electronics, except that the channel of a transistor was as small as the length of a single molecule, hundreds of times smaller than the state of the art for electronics. The picture was naive, but it seemed supported by Schön's data, as Schön included some of the same transistor curves as he had shown in earlier reports on organic crystal field-effect transistors. On May 4, Schön and Bao sent a manuscript to *Nature* containing the duplicated transistor data, claiming the invention of a "molecular-scale" transistor.

The review process for the paper (discussed in Chapter 6) gradually played out between Schön, Bao, and the journal over the next few months. At the same time, the paper came under internal scrutiny from other MTSs. On May 30, 2001, everyone in Cherry Murray's Physical Sciences and Engineering Division was invited to a seminar chaired by John Rogers and given by Schön, who had finally arrived back from Konstanz.

Schön introduced his new device by showing a cartoon. The cartoon showed a step made of silicon and coated with a layer of silicon oxide. On the step was a neat layer of organic molecules, standing up like blades of grass on a lawn. This was the transistor channel, with metal contacts below and above, which supposedly acted as the source and drain electrodes. Just as in the manuscript he had sent to *Nature*, Schön claimed to have applied a voltage to the silicon step, which acted as the gate electrode and projected an electric field onto the molecules, with the coating of silicon oxide acting as the gate insulator. Schön's data showed that the current switched in the molecular layer exactly as would be expected for a working transistor.

This "self-assembling monolayer field-effect transistor" was called the SAMFET for short. But not long into the seminar, Peter Ho interrupted. Ho started out by mentioning a couple of concerns that were obvious to many in the audience. Although the data Schön showed looked very typical for a conventional transistor, the SAMFET was anything but ordinary, Ho pointed out. In a conventional transistor, the distance from the surface of the semiconductor to the gate electrode has to be shorter than the channel; the distance between the source and drain electrodes. Otherwise, the effect of the gate field will be swamped by the influence of the source or drain, and no switching will take place. But in the cartoon graphic Schön showed, the molecules making up the channel were only two nanometers long, while the gate electrode that was supposed to switch them on and off was thirty nanometers away. Ho argued then that it did not make sense to use the concept of "mobility," which applies to the transport of charge in a conventional transistor, for such an unconventional design. This point was independently worrying a reviewer at *Nature,* and Schön was obliged to remove the discussion of mobility from the paper prior to publication. But after he spoke, Ho was responded to not by Schön but by Federico Capasso, who defended Schön, saying that mobility could be used as a rough measure of device performance even if it did not rigorously apply, a point that Schön and Bao made in a later paper.[14] Ho kept asking questions. Why did the data look so unlikely for the device? Dick Slusher, the director of Optical Physics Research, also spoke up to ask for details of how Schön had made his device. But by the end of his presentation, Schön did not seem to have answered the questions about why his data looked the way it did, and many MTSs and managers left the seminar puzzled. It seemed clear that some unknown effect was at work, which included the possibility of experimental artifacts that wouldn't have technological relevance. "No one doubted the data," explained Don Hamann, the director of Theoretical Materials Research, "they doubted the device was working by the mechanisms claimed."

Usually, scientists resolve this kind of uncertainty through discussion and additional measurements. But the circulation of the first SAMFET data took place against a background of deterioration in both management-staff relations and the quality of scientific debate. As Lucent's share price and revenues fell during 2001, the conditions for research worsened. To save money, building managers unscrewed many of the bulbs in the research buildings so that places where researchers might bump into each other and talk things through were less than half as well lit as they had been. Researchers had been able to spend up to $1,000 or $2,000 on research without management approval,[15] but as money became tighter, purchases as simple as paper towels for wiping up spills in the lab had to be ordered in bulk through department directors. This kind of restriction not only slowed the pace of research, it led to less respect for managers. Joke memos began to circulate that made fun of the cost-cutting measures. One suggested that scientists might sleep under bridges when they went to conferences instead of spending money on hotel rooms, while another forbade them from printing out papers using fonts greater than 12 pt without preapproval from management. Many staff began to feel that the essence of Bell Labs was under threat. When word spread that the Bell Labs library was slated for closure, one MTS, Mike Andrews, collected about one hundred email complaints before binding them up in a book and walking over to the office of Lucent's chief executive officer to hand them in. Andrews said one or more of the complaints quoted the Bell Labs motto, "off the beaten track," as an analogy to being able to browse in the library. After that, the library closure was delayed for a time, but Andrews did not receive a formal reply to his petition and was told informally that senior managers believed he had wasted time that he should have spent on research.

In July 2001, Capasso wrote to the lab. He explained that he knew the cost-cutting was slowing down research, but that there was nothing he could do. Lucent was "burning cash at an unacceptably fast rate" he explained. He said that he deplored the poor taste of a recent joke memo and added ominously that he hoped Lucent security would "zero in" on the sender. Capasso appeared to recognize that the quality of scientific discussion wasn't functioning very well and added that, with Lucent's budget too stretched to buy in food for meetings, he was going to pay for department pizza lunches out of his own pocket. The good news, said Capasso, was that at least the lab was still rolling out superb new science and technology.

But even as managers tried to sound upbeat, Bell Labs scientists were losing their jobs. Lucent offered early retirement packages to older researchers, and there were lay-offs at the spin-off company, Agere. The number of MTSs in the Physical Research Lab had fallen from 114 in 1997 to 56 in 2001[16] and people were anxious about jobs being cut in the future. So were managers, according to Steve Simon, director of Theoretical Physics. "People were getting fired, there was downsizing. We wanted to maintain some semblance of activity. It wasn't like anyone put the screws on and said 'publish more,' but technical problems with one person's data were not a big enough issue to get on the radar screen."

Then came the events of September 11, 2001. With Murray Hill within commuting distance of New York City, the disaster and its psychological aftermath felt close to home. From one of the lookouts on regular walks that MTSs could take around the wooded Murray Hill grounds after lunch, the plume of smoke from the ruins of the World Trade Center was clearly visible. One MTS told me that in this period, conversations in the cafeteria revolved around the world situation, the company stock price, and managers' intentions relative to that, and that this began to feel tiring, "but we couldn't stop." Another said that although MTSs had traditionally worked late at Murray Hill, he began to notice colleagues going home earlier because the days were getting shorter, and the research buildings and car park did not feel so safe during the evenings with some of the lights out. In October, Lucent's chief executive officer, Richard McGinn, was fired by Lucent's board, who brought back Henry Schacht, who had been CEO prior to the excesses of the dot-com boom. There were changes all down the management line. Arun Netravali, the Bell Labs president, left, and his position was combined with that of Lucent's vice president for corporate strategy, William O'Shea, who had spoken positively to a reporter about the relevance of Schön's work on the "Plastic Fantastic" in March. Bill Brinkman, who had been a particularly strong advocate for science, took retirement from his position as vice president for research, and was replaced by Jeff Jaffe, the manager who had come on board from IBM.

All the while, Schön's nanotechnology manuscript was making its way through two rounds of review at *Nature.* Over the summer, the Department of Condensed Matter Physics was renamed "Nanotechnology Research." As not everyone in the department worked on research that could be called nanotechnology, John Rogers joked to Julia Hsu, one MTS in the department, that at least everyone worked at a scale of at least "one million nano," a distance of a millimeter, Hsu said. Rogers's new title was "director of Nanotechnology Research," but he did not seem comfortable with it. Apart from jokes at the time, several of Rogers's later resumes and biographies stated that he had been "director of Condensed Matter Physics" throughout his time as a Bell Labs manager. With the publication of Hendrik Schön's paper in *Nature* in October, the Nanotechnology Research Department was baptized in a blaze of media coverage. Bell Labs sent out a press release claiming to have achieved a milestone on the path toward the creation of molecular-scale computing, the exact claim that a reviewer of the paper had tried to insist should be deleted from the scientific manuscript as a condition of publication, but that Schön had gotten away with including, apparently in part by mentioning the preliminary results under submission at *Science.* The Lucent press office organized a conference call for reporters, and Rogers was quoted as saying that the use of self-assembling molecules showed that the limit of lithography could be overcome, and that the prospect of industrial-scale manufacture "looks very promising." A broadcast of the conference call was saved on the Lucent web site under the title "Bell Labs Product Announcement." Although the reference to a "product announcement" might have been an error, it only helped to further the impression that Schön's work was providing immediate value to the company.[17] In the press release, Cherry Murray explained what the relationship of Schön's work to the company's bottom line might be "This work shows the value of long-term research. Although there may be no practical applications for a decade, it could lead to a new paradigm in electronics." She added that there were gaps in the theoretical understanding of the devices, apparently revealing her awareness of discussions in her division that had included consideration of the possibility that Schön was measuring an artifact, so that his devices were not, after all, technologically relevant. Finally Federico Capasso invoked Murray Hill's history as the birthplace of the transistor with the comment "the molecular-scale transistors that we have developed may very well serve as the historical 'bookend' to the transistor legacy started by Bell Labs in 1947."[18]

GOOD TEAM PLAYERS

After the publication and press release of Schön's SAMFET by managers, opinions rapidly became more polarized. The Murray Hill cafeteria had once been a place of vigorous scientific discussion between researchers with diverse views, but the spin-off of Agere pushed people to sit with like-minded colleagues more than they might otherwise have done. One regular group of eaters included Kleiman and Willett, two MTSs who were expert in electrical measurements on field-effect transistors, and David Muller, who had asked to image Schön's laser samples in an electron microscope the previous year but who had never been able to get samples. They were in Federico Capasso's lab, but still hung out with several researchers who had been spun off to Agere including Don Monroe and Jack Hergenrother, who had expertise in making ultrasmall field-effect transistors. The consensus in this group was that Schön's SAMFET could not be as described. The group was very concerned by the fact that the gate electrode was thirty nanometers away from the transistor channel, too far to modulate current strongly in a channel that was only a couple of nanometers long. Although there was a possibility that, as Cherry Murray had claimed in the press release, some new theory was needed, it seemed more likely to experts who had made or characterized transistors that Schön had made an elementary experimental mistake. Some of the possibilities were as follows. Willett thought Schön might be getting some of the gold for the electrodes into the channel of the transistor, producing a short circuit. Monroe thought Schön might have formed a "parasitic transistor," a transistor in a different part of his sample than he thought, so that the true dimensions of his device were different from those he had reported. Muller wondered if Schön's gate insulator was thinner than he thought, something that could have been established by looking at the samples in his electron microscope. But no single explanation emerged as the obviously correct one, and Kleiman said the group racked its brains repeatedly. "We thought it wasn't right scientifically. We said, 'what mistake could he have made that could cause him to measure this?'"

Meanwhile managers and collaborators of Schön, including John Rogers, Federico Capasso, Zhenan Bao, Christian Kloc, Horst Störmer[19] and Zhenan Bao, the chemist who was working with Schön on the SAMFETs, tended not to eat with the group in which the most skeptical views were emerging. This was not a deliberate segregation, but it could have had a significant effect, given that much of the scientific discussion at Bell Labs was understood to take place in just such informal settings. Schön assiduously avoided sitting with either group, hanging out instead with German-speaking postdocs who weren't involved in his research, leaving others to fight his battles for him. One lunchtime, for example, Kleiman got into a discussion with Steve Simon, the director of Theoretical Physics Research, who was interested in trying to understand Schön's effects and often reached out to the critics. Kleiman argued that Schön shouldn't have published the single molecule transistor claims without including a proper interpretation of how and why the device worked, or saying openly in his paper that the results did not make any sense. When data is, on the face of it, implausible, the burden of proof ought to be on the experimentalist to explain how it came about, said Kleiman. In response, Simon argued that there was nothing wrong with an experimentalist publishing remarkable results in the absence of an explanation for the underlying mechanism that explained them, as further work, for example by theorists, might help to provide one. In a way, the unusual nature of Schön's results only made them more interesting. Kleiman said that although the argument was heated, in hindsight he credited Simon with being a manager who was making an effort to talk with critics of the Schön work. Eric Isaacs, who had managed Schön in late 2000 and who was Kleiman's director, had told Kleiman that some managers believed he and others critical of Schön's work were only "jealous" of the young MTSs' success, Kleiman said. Don Monroe, who was at Agere, and so somewhat independent of the dynamic inside Capasso's lab, said that the feeling among the skeptical Bell Labs staff he knew was not so much jealousy but disappointment that through the high-profile publication and press release, managers had tied the reputation of Bell Labs to research that was not well understood, and so might turn out to be wrong.

Following the spin-off of Agere in 2000, Lucent had retained a 57.8 percent stake, but steps were under way to sell this. Resources for research became even tighter as Agere ownership stickers started to appear on lab equipment. Julia Hsu, who had access to a $300,000 atomic force microscope for imaging samples, discovered that the instrument was going to Agere. She pointed this out to her manager, John Rogers, who told her not to complain. Hsu said Rogers suggested she take a look at Hendrik Schön, who was producing great results while working in someone else's lab using hardly any resources. By this point, it was widely known that Schön was doing much of his work in Ernst Bucher's lab in Konstanz. Schön was very inexpensive, having placed the bill for the molecules he used in his SAMFET devices on Christian Kloc's tab, even though Kloc was not a collaborator on the work. Schön was also reluctant to apply for a corporate American Express card to buy and track expenses for his own lab until Kloc insisted. In a time of scarce resources, Schön must have seemed like a manager's dream. But Julia Hsu pointed out that it was not possible to do good science for free. Someone had paid to set up Schön's equipment in Konstanz, even if it wasn't Lucent, she argued.

On October 31, 2001, Schön and Bao gave a second division seminar, chaired by Bao's department director Elsa Reichmanis. This time, Schön got so many questions that at one point, Federico Capasso intervened forcefully and asked everyone to let Schön get on with his talk, making a comment that one MTS wrote down as, "just because you don't understand the science, doesn't mean it's wrong." Seminars at Bell Labs had always been famous for an atmosphere of debate, in which challenges to the speaker were much encouraged. The combination of Capasso's enthusiasm for Schön's results, with his brusque manner in trying to move things on, created the impression among some MTSs that managers were reluctant to see Schön questioned. Chandra Varma explained to me that managers' reaction to questions about the SAMFET data had been that Schön should be allowed to continue, to explain things better, so that further research could clarify how his devices worked. At this seminar, Schön also experienced a boost from the data of his collaborator, Zhenan Bao, who claimed to have constructed a device similar to his own. Then Kleiman spoke up with an objection. Bao had claimed that at least two different SAMFETs worked by a similar physical mechanism. Kleiman pointed out that the width of the channel in one of the devices was twenty-five micrometers (a micrometer is one millionth of a meter) while in the other, it was less than one micrometer. With such different dimensions involved, the physical mechanisms had to be different, Kleiman said. Schön responded for Bao, telling Kleiman that the larger distance wasn't twenty-five micrometers, but five. Kleiman was surprised because he remembered seeing the larger number written in a copy of Schön and Bao's paper. After the seminar, Kleiman looked up the paper, which was circulating as a preprint internally and was under submission at the journal *Applied Physics Letters,* where it later appeared.[20] The distance was twenty-five micrometers, as Kleiman had thought. Although in principle it was possible that Schön might have forgotten what his own paper said, Kleiman reached a different conclusion. "He was lying," he said, figuring that the misstatement had enabled Schön to sidestep a confrontation over the operating mechanisms of the SAMFETs. But, even thinking this, Kleiman still did not realize that Schön had actually faked the data.[21]

Kleiman was not the only MTS to come out of the October seminar feeling uneasy. During Schön's presentation, Julia Hsu noticed an image of the devices taken using an atomic force microscope. After the seminar, she approached Schön to ask which instrument he had used. Schön told her that this was not his data, but came "from the literature." Hsu was taken aback. Her impression from the presentation was that this was an image of Schön's device, taken by Schön or perhaps a collaborator, as Schön had not provided any citation or credit to an outside group's work. Hsu did not think that this was plagiarism or misconduct, but she did think that it was sloppy and not the right way to present data. "I remember thinking it was misrepresentation," she told me. She then reasoned that because this was only an internal seminar, the issue might not be regarded as very serious if she spoke up.[22] Hsu did not know that Bao, and later Schön, had been asked for images of the devices once already by Peter Ho, that Schön and Bao had been pressed again by editors at *Nature.* Hsu also had some reason to try not to say anything that might be perceived as a complaint about Schön. In her performance review, in September 2001, Hsu was given a much lower score than she expected on the basis of her scientific work. When Hsu asked Rogers, her manager, why, he explained that she had been downgraded for not being a good team player, because she had complained about resources, an apparent reference to the discussion about Schön and the atomic force microscope going to Agere, she said. In the end, Hsu did not mention Schön's unexpected confession to anyone, not to colleagues, not to friends, and not to Bell Labs management.

"IT'S GOOD TO BE BRAVE"

When Ernst Bucher, Schön's former PhD supervisor, visited Murray Hill in 2001, he found himself called unexpectedly into Federico Capasso's office. There, he said, he found that Capasso and Rogers wanted to talk to him about Schön. Bucher said the managers were concerned about the reproducibility of Schön's postdoc work in organic electronics. They asked Bucher if he would speak to Schön and use some of his influence as Schön's PhD supervisor to encourage his former student to slow down the pace of his claims and help other people to replicate them. Bucher later spoke to Schön, but, he said, he did not feel that he got through.

Bucher said he felt that the managers approached him because of the possibility that he might be able to speak to Schön gently, without upsetting him. If Bell Labs managers had acted too confrontationally toward Schön, they were afraid he might leave, depriving the lab of one of its most exciting up-and-coming young scientists. At the same time, the managers were acutely aware of the reproducibility problems; by around the spring of 2002, Theo Siegrist and others had suggested to Federico Capasso that Lucent might want to hire Bruce Van Dover, who had expertise in sputtering but who had been spun off to Agere, to assist with setting up sputtering equipment at Murray Hill. An additional perspective to take under consideration is that people dealing with Schön often experienced difficulty addressing things with him. Schön did not respond well to confrontation, but he could be exceptionally prolific in response to any suggestion, causing others to feel cautious about triggering him in the wrong direction and perhaps making a problem worse. Some colleagues also reported a visceral reaction of aversion as they began to understand how unusual Schön was. His relentless supply of data was matched by an equally uncanny reluctance to engage in discussion.

There also seemed to be a reasonable chance that problems in Schön's work might be solved by involving MTSs other than Schön. For example Chandra Varma was working to understand Schön's work, as he would have done for any new results coming out of the lab, he told me. Varma came up with the idea that Schön's gold electrode might contain an imperfection that acted like a knife edge, massively focusing the electric field from the gate; this was an effect that could open up huge possibilities for beating the physical limits usually thought to constrain the properties of transistors, and perhaps also explain the data from the SAMFET. Two staff who had been working on molecular electronics before Schön, Nikolai Zhitenev and David Abusch-Magder, also became involved through an informal working group that met to discuss all aspects of molecular electronics, including Schön's results. Another idea was that the organic molecules in the SAMFET changed their shape in response to the subtle effect of the distant gate electrode, radically changing their properties and producing a change in current far larger than might otherwise be expected. The feeling among experts internally was that because molecular electronics was a new field, it was to be expected that devices might behave in unexplained ways. On the outside, Tom Jackson of Pennsylvania State University agreed, approving Schön's second nanotechnology manuscript for publication in *Science* after asking that more discussion of possible explanations for his data be included, including information about the possibility of the molecules changing shape. Jackson said he found Schön's data puzzling, but that other important claims in the field of molecular electronics, such as those from the Yale and Hewlett-Packard groups, were also in need of clarification. Mark Reed at Yale said that it was clear from the observation that Schön's gate insulator was much thicker than the length of his transistor channel that Schön could not have obtained his results in the experimental configuration he described, but that this only made it a more interesting challenge to try to figure out what Schön might have measured instead.

As Schön's second paper came out in *Science,* Bell Labs managers issued a second press release about the SAMFET. This coincided with a Lucent investor meeting in New York City given by William O'Shea, the new Bell Labs president and the Lucent vice president for corporate strategy. At the meeting, O'Shea gave both Schön's plastic superconductor and his molecular transistor as two of five "disruptive" technologies that might change the face of future technology and that had been developed at Bell Labs. The claim was broadcast on the Internet and was made part of an official report to investors filed by Lucent at the U.S. Securities and Exchange Commission.[23]

As Schön's profile kept rising, Bell Labs managers entered him for various prize competitions. In 2001 they learned that he had won the Materials Research Society's Young Investigator's Prize for 2002. He was also in contention for the TR100, a list of Young Innovators publicized by the Massachusetts Institute of Technology's *Technology Review* magazine, and was later selected. In November 2001, he was invited onto Ira Flatow's *Science Friday,* the weekly science phone-in broadcast by U.S. National Public Radio, where he fielded questions from callers on the SAMFET. On the show, he proved as reluctant to get into confrontations as he was with colleagues at Murray Hill. He stopped just short of contradicting one caller who asked whether, maybe, because his molecules were organic, they might be interfaced with the human nervous system. He endorsed the Flatow's suggestion that the devices might be useful in responding to chemical warfare, for example as a sensor. Flatow declared himself "blown away" and his brain "overloaded" by what Schön was claiming before wrapping up the segment by saying how interesting it was that all this was happening at the place where the transistor had been invented in 1947.

So while managers were, in principle, trying to get Schön to slow down, they were sending out very mixed signals, which barely impinged on Schön's steadfast commitment to the imperative of trying to publish as much fabricated data as possible. By the end of 2001 he had slowed down submissions somewhat but was still publishing at a faster rate than any other MTS at Murray Hill,[24] rising to an extraordinary peak of seven papers in November 2001. Although his interpretation of his data was increasingly under fire, he took refuge in the possibility that further research might help out, as well as the idea that in some ways, the unexpected nature of his results only made them more interesting. Striking out on his own as a star was not only enjoyable; he could find justification for his behavior in the expectations of others. Speaking to a reporter for the local paper, the *Star Ledger*, Schön explained, "Someone told me once, 'It's good to be brave. Challenge accepted wisdom.' So I've been trying to do that."[25] Gone was the cautious trainee who had worked so hard to reach agreement with other scientists. As a star, Hendrik Schön had found a way to rationalize the unexpected too.

10

THE FRAUD TABOO

A whistleblower of scientific fraud once told me that he felt he needed the right to remain anonymous for the rest of his life. "Like a rape victim," he said.

To begin with, I thought the analogy overblown. However, over time I heard so many other scientists struggle to articulate their strong aversion to being associated with fraud that I began to take it more seriously. Thinking of a fraud allegation as if it were an allegation of sexual abuse, I could start to understand on an instinctive level why scientists might feel strongly and yet be very fearful about coming forward.

Steven Shapin, a historian of seventeenth-century science, has traced the origin of the taboo on fraud back to the sensibilities of the gentlemen who founded the Royal Society in England in 1660, when calling another man a liar was likely to lead to a dueling challenge, and so was a question of life or death.[1] Understandably, early scientists were extremely loath to question the observations of others, and any situation in which different observers seemed to be in conflict had to be approached with enormous care. Shapin documented one incident in 1665 in which the observations of two astronomers disagreed in such a way that one of the two had to have made an incompetent error. This produced a crisis that led to the formation of a group of distinguished Royal Society astronomers to arbitrate. Had either man been accused of lying, the crisis would have been even worse. Harry Collins, a sociologist of science who has won grudging respect from physicists for his careful field studies of researchers involved in the search for ripples in space-time known as gravitational waves, later speculated that scientists' aversion to strong allegations is linked to the expectation that others conduct themselves cautiously in making claims.[2] Collins found Anglo-Saxon researchers less likely than continental Europeans to approve of bold claims to have detected gravitational waves on the basis of tentative signals; he attributed this to the possibility that they were closer to the traditions of the Royal Society. Because they felt limited in their ability to question observations, Anglo-Saxons held their colleagues to high standards of rigor in the interpretation and staking of new claims, Collins speculated. In contrast, Europeans were apparently more comfortable both with bold scientific claims and questions about others' observations and motives.

Collins's idea was speculative, but looking back in the history of scientific fraud, it is true that some of the first public fraud allegations took place in continental Europe, at a time when journal editors were publishing some very provocative reports. One of the first whistleblowers I came across was Georges-Louis Le Sage, a Geneva-based teacher of mathematics and physics who went public with a fraud allegation in 1773. The historian and physicist James Evans summarized the case in English in the journal *Isis* in 1996.[3]

In 1798, Evans wrote, the Paris-based *Journal des beaux-arts et sciences* published a letter from a physicist by the name of Jean Coultaud, who lived in the French Alps not far from Le Sage's home city of Geneva. Coultaud described himself as a former professor of physics from the University of Turin in Italy. He claimed to have run a pendulum experiment in Samoëns in the French Alps, placing two of Geneva's finest pendulum clocks on a mountain, one about 2100 meters above the other, and leaving them for two months. When he reunited the clocks, he said, he found that the higher one was running half an hour ahead. This contradicted Newton's theory of gravity, which predicted that the clock nearer to the center of the Earth should feel a stronger gravitational field and so tick faster. Coultaud claimed to be surprised and to have run all kinds of careful cross-checks, including one that found that the pressure of air was the same at the two altitudes, so that variation in this quantity could not explain the results. The editor of the *Journal des beaux-arts et sciences* put Coultaud's article in the lead position, the eighteenth-century equivalent of getting on the cover.

Fascinated to read about the apparent refutation of Newton, Georges-Louis Le Sage asked a mountaineering friend, Jean-Andre Deluc, for help contacting the new local physicist. But Deluc was the inventor of a barometer, and knew, from his own travels in the French Alps, that the higher he went, the less dense the air, an observation in contradiction with Coultaud's claim to have found that the density of air did not vary with altitude. At Le Sage's suggestion, Deluc traveled to Samoëns, where Coultaud claimed to live, but he returned to Geneva to tell Le Sage that no one in the area seemed to have heard of the pendulum-clock physicist. Puzzled, Le Sage visited clockmakers in Geneva, but no one had placed an order for two large, identical pendulum clocks. Finally, Le Sage wrote to a professor at the University of Turin, Italy, who said he had never heard of Jean Coultaud.

Le Sage's next step was to contact a prominent anti-Newtonian scholar in Paris, Joseph-Etienne Bertier, to express concerns that the Coultaud report was bogus. It's not clear why Le Sage did this, but he may have been trying to resolve the matter privately with scholars in the anti-Newtonian camp before going public. If so, he suffered for his caution, as Bertier retaliated, publishing a pre-emptive attack accusing Le Sage of spreading unsubstantiated rumors.

Le Sage was forced to go public. In 1773 he published a fraud allegation in the journal *Observations sur la physique,* describing his problems in locating Coultaud, and irregularities that he and Deluc had noticed in his observations. He also identified a second Alpine experimentalist who had published in *Journal des beaux-arts et sciences* and claimed to replicate Coultaud's finding, but who could not be traced either. Although today scientists like to say that fraud will come to light when other scientists fail to repeat work, Le Sage had stumbled across an extraordinary example of not only an original report, but also a replication effort, apparently being faked. After publication, neither physicist came forward, and *Journal des beaux-arts et sciences* published a statement admitting that its editor had been duped. Although the author of the two pseudonymous reports was never found, Evans suggested in 1996 that Hyacinthe-Sigismond Gerdil, a cardinal in the Catholic Church, was the culprit. Like Coultaud, Gerdil was a former professor of physics from Turin, had been born and raised in the village of Samoëns, had enough mathematical sophistication to have authored the articles, and had a bruised ego because attempts he had made to refute Newton under his own name had failed.

Like many modern-day whistleblowers, Le Sage worried about trying to do the right thing, and he suffered retaliation. His exposure of the Coultaud fraud was also arguably helpful to the progress of science in France, as Coultaud's claims had sparked several replication attempts. Le Sage continued to point out errors and confusions in experiments on gravitation for years after exposing the Coultaud report.

He was followed by other European whistleblowers. In 1820, the German physicist Johann Franz Encke published an accusation of fraud against Jean Auguste d'Angos, the director of an observatory in Malta, in *Correspondence astronomique*, a journal published by the Hungarian baron Franz Xaver von Zach. D'Angos had claimed to have discovered a comet in 1784, but had been under suspicion since 1806, when he had told an astronomer who asked for some of his observations that he had lost his records in a 1789 fire at the observatory, without mentioning that he had published at least some of the data in 1786 in the *Leipziger Magazin fuer die Mathematik,* where Encke later stumbled across them. Encke was able to show that the calculations fit exactly what would be expected if d'Angos had calculated them using an erroneous parameter.[4]

By 1820, at least two whistleblowers of fraud in Europe had been vindicated. But in 1824, Zach's *Correspondence astronomique* published a fraud allegation that turned out to be far more problematic. The accusation relied on information from a former worker at an observatory in Buda, Hungary, Daniel Kmeth, who could not reconcile raw data from the observatory with publications by Johann Pasquich, the observatory's director. Pasquich was a very highly regarded astronomer, and the accusation against him provoked the formation of a panel of five prominent astronomers to investigate. The panel found that they could replicate Pasquich's published orbits from Kmeth's data, which Kmeth had included as part of his published allegation.

They published an analysis called "the saving of Pasquich's honor" in the *Astronomische Nachrichten* in 1824. Both Kmeth and Pasquich had been harmed by the scandal.[5]

This, the risk of being wrong, is frequently cited as the major problem with making a public fraud allegation. Evaluating a situation correctly can be especially difficult when the people involved have previously worked together and are relying on previously unseen evidence when they go public. In America today, scientists' view of fraud allegations has been profoundly influenced by the legacy of a late twentieth-century case that fits this description. This is the so-called Baltimore case, named for David Baltimore, who won a Nobel Prize for Medicine in 1975 at the age of thirty-seven, and in 2008 was still one of the most influential biomedical scientists in the United States. Historians have disagreed about the evidence in this case,[6] but have agreed that it has had a far-reaching legacy. This is true even in physics, with which Baltimore has had little personal involvement.

In 1986, Baltimore and several coauthors published an apparently significant paper in the journal *Cell.*[7] Data in the paper were subsequently questioned by a research scientist who had been assigned to follow up on the work. Baltimore was not involved in the collection of the questioned data and was not the target of the accusations, but as a coauthor on the paper he became involved in defending its key claims. MIT's ombudsman tried to arbitrate, but complaints about the process reached a congressional subcommittee headed by Democratic Congressman John Dingell, who aggressively pursued the case, citing concerns about the vast amount of money American taxpayers spend on science at universities. The resulting debacle proved so embarrassing that Baltimore resigned from a new appointment as president of Rockefeller University in New York City after only one year in the job. His coauthor was found guilty by an investigation convened by the U.S. Department of Health and Human Services' Office of Research Integrity, but cleared on appeal in 1996, ten years after the publication of the original paper.[8] With her exoneration, Baltimore's defensive stand was reinterpreted by many scientists as heroic. He became president of the California Institute of Technology in Pasadena, and in 2006 was elected president of the American Association for the Advancement of Science.

The message that many scientists took away from this case is that it is risky and unethical to go public with a fraud allegation because of the possibility of smearing an innocent person and of provoking a political backlash against science. The *New York Times* has even archived its articles about the Baltimore case under the heading, "Science on Trial."[9]

In August 2005, I encountered the legacy of the Baltimore case while following up on rumors about possible fraud by Luk van Parijs, an immunology professor at the Massachusetts Institute of Technology. To get a feeling for whether there was anything in the rumors, I had looked through some of Van Parijs' papers and noticed a pattern of apparent irregularities, instances in which Van Parijs had duplicated and miscaptioned some of his data. This had no direct relevance to the Baltimore case, but the resonances were impossible to avoid; Van Parijs was at MIT, where Baltimore and his accused coauthor had previously been based, and had been a postdoc in Baltimore's lab at Caltech. Baltimore was senior author and corresponding author on two of the papers containing irregular data.

Before publishing an article, I sought advice from immunologists on the evidence I had discovered, and found them so reluctant to help that I began to make a spreadsheet of their reasons for declining. Four did not respond to my inquiry. Three said they had to go to a scientific meeting. Three said they did not have any information about the people involved. Two said this was not their field. Two said their employers would not allow them to comment on misconduct. Two said getting involved might affect their relationships with colleagues. One said simply that he was unable to help. One said he had to concentrate on a grant application. One said he had been involved with misconduct before and could not face it again. One said his fiancée had recently left him. One said he would not become part of a kangaroo court against a colleague, and one (Michel Nussenzweig of Rockefeller University and the Howard Hughes Medical Institute) said that he did not understand why I needed to ask him, when I could ask a student or technician to comment instead. Of seven immunologists who agreed to look at the problematic data, three were prepared to be quoted. My work led to an inquiry by Caltech and to a misconduct finding against Van Parijs in 2007.[10] By that time, word of the evidence had spread, and Van Parijs had been fired over separate allegations of misconduct brought by junior researchers in his MIT laboratory.[11]

Because of the Baltimore case, immunologists might be expected to be more sensitive about inquiries into alleged fraud than other scientists, and I have found physicists more open. But physicists also suffer anxiety about the possible repercussions of public accusations, and on two occasions even physicists have brought up the Baltimore case to me as an example of how things can go wrong. The U.S. Federal Policy on Misconduct, drafted in 1999 following the resolution of the Baltimore case, enshrined scientists' anxiety about false allegations in regulations that say allegations should be investigated both confidentially and thoroughly by the responsible institution before anything is made public. The Office of Research Integrity, which has oversight of science funded by the U.S. National Institutes of Health, adds to this a warning that reporters can be considered "unauthorized persons" when it comes to the discussion of suspected fraud.[12]

But suppose Harry Collins was right that scientists' preference for cautious science is culturally connected to their aversion to participating in fraud allegations. Then it should not be a surprise to find that scientists more likely to make allegations, and to make them in public, when they are confronted with bold, high-profile claims. This is the case even though, in principle, both low-profile and high-profile science could be fraudulent. Today, journals such as *Nature* and *Science,* and even scientific societies like the American Physical Society, are actively engaged in promoting research claims, as are the press offices at most research institutions. The increased profile can create a perception, or a reality, that institutional officials will not run a confidential investigation objectively and will not report the findings publicly, increasing the likelihood that accusers feel they have to go public if they want to warn others about the irregular research. And for important scientific claims, the stakes become high for inaction as well as for action; high-profile fraudulent research can divert resources from authentic work or do long-term harm to people or fields of research that are later found not to have taken action when they could.

So when and how should whistleblowers go public? Although many institutions have misconduct policies, those policies rarely advise scientists on how to balance confidentiality with the obligation to let the public know that high-profile claims have fallen under suspicion. In addition, policies rarely tell scientists when to make allegations, but only stipulate how an institution should respond when they do. In part because of the taboo on fraud, scientists with suspicions sometimes fail to identify themselves as "whistleblowers" right away, instead feeling that they are only critics with concerns, leaving it up to officials to make the call on whether or not an investigation is in order. None of this means that today's scientists are not prepared to challenge the taboo on fraud and bring forward allegations of fraud, either privately to officials, or publicly when they think it is necessary, but it does mean that both whistleblowers and the institutional officials responding to them are often left with little to go on but their instincts.[13]

TOO GOOD TO BE TRUE

The publication of Hendrik Schön's SAMFET paper in *Nature* came promptly to the attention of Don Monroe, one of the Bell Labs MTSs who had been assigned to Lucent's spin-off company, Agere Systems. Because Monroe was no longer part of Bell Labs, he had not been invited to the division seminar in May 2001, and his last memory of Schön's work was from a Bell Labs talk that Schön had given in 2000. For Monroe, the talk had been tainted by a sense of nostalgia. Going with Agere, Monroe realized that he would lose touch with much of the fundamental science going on at Lucent. Schön's work seemed like an example of the kind of exciting cutting-edge research that had made Bell Labs into such a wonderful place to work. There things stood until October 2001, when Schön's SAMFET paper appeared in print, the lab sent out press releases, and Monroe became one of several Agere scientists involved in discussions about Schön's work again. Monroe talked and emailed with researchers at IBM Research, which was traditionally competitive with Bell Labs, and with a colleague who had been asked to comment for an article by a *Wall Street Journal* reporter. Despite the spin-off, many scientists inside and outside Bell Labs still considered Agere scientists, who remained located at Murray Hill, as part of the Bell Labs community. Monroe also had a special expertise in the design of ultrasmall transistors. In 1999, Monroe, his colleague Jack Hergenrother (also spun off to Agere) and others had reported a way of building a silicon transistor on its side, laying down one layer of atoms one at a time, rather than by using optical lithography to etch patterns onto the horizontal surface of a piece of silicon. Schön had copied a sentence from their report into two of his own manuscripts, and even suggested that the SAMFET worked in a similar way.[14]

Former colleagues described Monroe as a very methodical thinker. The way Monroe put it, his second thoughts on any topic were usually a better guide than his first. His first thought, after the SAMFET appeared, was that it was possible that all the questions being asked about Schön's device arose from a miscommunication about how it had been made, which, as someone with related expertise, Monroe felt he might be able to clear up. One of the questions arising at Agere was why, as Schön switched his gate electrode on, the rate of increase in current was not only greater than would be expected given the large distance from the gate electrode to the transistor channel—this issue had already been noticed both by a reviewer for *Nature* and by MTSs at Lucent in May—but was even greater than a physical limit imposed by the laws of thermodynamics, which describe how fast electrons jiggle in the electrodes of a transistor at a given temperature. Transistors switch on as the gate voltage lowers the energy needed for electrons to spill out of the electrodes into the channel, so the devices cannot switch on faster than the rate of jiggling. As Monroe knew from his own experience, researchers building new transistors can hope to approach the thermodynamic limit, but only ever pass it in exceptional circumstances.

Monroe started out by asking Schön's coauthor, Zhenan Bao, about this issue. She did not have any answers and referred him to Schön, who admitted that he couldn't explain the operation of the devices either. "Even our theorists could not come up with any good explanations," he said, using "our" to invoke his affiliation with Bell Labs. Monroe said he wanted to help. He told Schön that several people at Agere were interested in the work and invited him to give an in-depth seminar. He took the precaution of warning that he couldn't promise a friendly audience but could promise a fair dose of critical thinking. In response, Schön said that he was already giving a division seminar at the end of October, although he wasn't sure if Agere people would be allowed to come, adding his characteristic ellipsis ". . ." to try to provoke interest. But Monroe was not distracted, and insisted on the idea of an in-depth seminar, hoping to get into more detail and ask tougher questions than he would feel able to do with the whole division watching. Schön agreed, and they arranged to meet on October 29, two days before Schön would talk to the division. On the day, both Monroe and Schön arrived early, and had a short conversation in the Agere seminar room a few doors down the corridor from Schön's lab. Monroe congratulated Schön on his streak of remarkable publications. Schön tried to play down his success, saying that he wasn't sure it would last.

Around half a dozen people arrived and the seminar began, with Schön showing data from both his recent *Nature* paper and a paper that was forthcoming in *Science.* Schön's presentation was smooth as he started off with a clean slate for the new audience and did not point out problems that they did not ask about first. When people began to interrupt with questions, pointing out ways in which the results did not make sense, Schön shrugged or agreed. "Anything we asked, he would say, 'this is what I measured, this is what I see,'" said Ashraf Alam, one of the Agere researchers who attended. As well as dozens of technical questions, there was one from Alam that Monroe found particularly unsettling. By this point, Schön had reported diluting his molecular layer to the point where some devices contained as few as three, two, or even just one molecule acting as the transistor channel. His evidence for this was a bar chart showing that the conductance, the ease with which current moved through each device, depended on how many molecules it contained—the greater the number of molecules, the larger the conductance. Alam pointed out that the outline of the chart seemed to have the beautifully smooth and symmetrical shape of a Gaussian, a bell-shaped curve that is often used by physicists for approximating quantities that they have not measured. Having run reliability tests on prototype devices, Alam knew that plots of their properties rarely fall on a perfect Gaussian, especially for the small number of devices, around thirty, that Schön claimed to have studied. He asked for an explanation, but Schön did not seem to understand and went into an elaborate description of how the chart had been derived from his original data that did not seem to make much sense.

Later, on the way back from the seminar, Alam suggested to Monroe that they run a basic statistical test, a chi-squared test, on Schön's data. A chi-squared test would reveal whether the results fit unrealistically closely to a Gaussian, as it appeared. Monroe did a rough analysis of Schön's data straight away, relying on a printout Schön had distributed at the seminar. Alam was right: the results were too good to be true. That evening, Schön sent a pleasant note thanking Monroe for the discussion. Monroe responded and continued to discuss a number of technical points, but he kept thinking about the issue of the statistics. He could double-check his rough analysis easily with access to the original data underlying Schön's bar chart, but all of a sudden, he didn't feel comfortable asking for it. "It was starting to cross the line into something that didn't look like it could be an honest mistake," he said.

In principle, this was the first time Monroe had doubted a colleague's honesty, although thinking back, he recalled exchanging a joke with Dimitri Antoniadis, a professor of electrical engineering at MIT. It was Antoniadis who had been asked to comment on Schön's SAMFET work by a *Wall Street Journal* reporter and struck up a discussion with Monroe. Feeling cautious, Antoniadis didn't comment in the end, but told Monroe that the device performances reported by Schön reminded him of the runner Rosie Ruiz. Ruiz had won the 1980 Boston Marathon, but she was later found to have cheated by registering and then jumping in from the crowd and sprinting to the finish line with all the cameras on her. The story was that she didn't mean to win, only to impress her friends by finishing with a good time, but got caught because she mistakenly finished first. Monroe laughed, but then corrected himself. Antoniadis was only making the analogy in the sense that, by claiming performance better than had been achieved in conventional transistors made from silicon or gallium arsenide, Schön might come under greater scrutiny than he expected. Ruiz had cheated, a different thing. But the incident may have opened a chink in Monroe's assumption that scientific colleagues always acted honestly. Over the couple of months that followed the SAMFET publication, there were other jokes about Schön's work circulating over lunch in the Murray Hill cafeteria. People joked about famous examples of "pathological science," a phenomenon described in a 1953 lecture given at General Electric Laboratories by the American physicist Irvin Langmuir. One of the hallmarks of "pathological science" is supposed to be that scientists get excited about a physical effect that is at the edge of what can be detected and begin to act uncritically in defending its reality against any criticism that comes up. "Pathological science" does not necessarily involve the kind of intentional data fabrication that Schön was engaged in, and that Monroe was beginning to suspect, but it is a general category of bad science that physicists sometimes joke about when they feel they are having difficulty in getting their criticisms to stick.[15]

Ten days after the seminar, Schön's paper containing the conductance data appeared on *Science Express, Science*'s mechanism for conveying "hot" papers to the scientific community.[16] Monroe went online, extracted numbers from the chart he was concerned about and put them into a spreadsheet to check his earlier analysis. It looked about right. He wondered whether some kind of wishful prejudice might have produced a distortion in the data. Could Schön have disregarded some data because some devices didn't work, subconsciously hoping to get a better final result. That would reduce the amount of noise in his final chart, without constituting intentional fraud. But no matter how hard he tried, Monroe couldn't find a way to explain the smoothness of the plot, even if the data had been selected in some way from measurements of real devices. With a shock, he realized that the only way to explain the data was that the devices had never existed in the first place.

Over these few days, Monroe mentioned his worry about overly good statistics to Steve Simon, a director in Federico Capasso's lab who sometimes ate lunch with MTSs who were critical of Schön's work.

Monroe also sent Schön a note, copying Zhenan Bao, in which he said that he was unable to think of an objective measurement process that could explain the data, and asked for a response. Schön replied to mention the possibility that he had plotted his data in such a way that some data points had mistakenly been hidden, but ended by agreeing that the question of statistics remained, which he punctuated with three trailing dots. Monroe did not follow up with Schön but contacted Schön's director, John Rogers. He sent Rogers a file containing his statistical analysis and met with Capasso to talk about the possibility that the data were not acquired properly. Capasso asked whether Monroe had some kind of "smoking gun" for what he was suggesting. Monroe did not come up with a good answer in the meeting, but he thought about it. On November 20 he wrote Capasso an email with the subject line "Smoking Guns" in which he said that he believed, based on the statistics, that there was a greater than 90 percent chance that Schön's data had been distorted by some kind of human factor, and that there was a greater than 50 percent chance this was intentional. Even though Monroe had phrased this carefully, most scientists would recognize his communication as an allegation of scientific fraud. Monroe added that he understood that this kind of thing probably couldn't be resolved through normal collegial interactions, and that managers might decide to open an "adversarial investigation." Apparently realizing that they might have difficulty going adversarial on Schön, he offered to help out if they needed a "bad cop."

Capasso's response was to thank Monroe, and to say that John Rogers had talked with Schön about the need to slow down the pace of his research claims. The managers had spent the few days between Monroe's first mention of the conductance data and his allegation documenting some concerns about the rigor and replicability of Schön's work. This was incorporated as part of Schön's performance review in November, somewhat late relative to the date that other former MTSs remember other performance reviews taking place, which was September/October. Managers were also trying to arrange for a replication of Schön's work that might be considered "independent," including considering whether Horst Störmer, the Nobel Prize–winning physicist at Columbia University who was collaborating with Schön under a grant from the National Science Foundation, would also be interested. They also asked Schön to hold off on the submission of a paper reporting on the invention of a single electron transistor, a transistor that could switch current one electron at a time and that built on the SAMFET work. None of this was directly related to Monroe's allegation, but implies that managers had broader worries about the reliability of Schön's work. In response to the allegation, Capasso said only that he would like to look into the matter and see results from other experiments that Schön might do.

Over the next few days, John Rogers and Federico Capasso looked into the issue raised by Monroe. They took slightly different approaches. As might be expected from Capasso's position as a more senior manager, Capasso grasped that the major issue was a question of research integrity, and was both realistic and unhappy about the fact that it wouldn't go away. Although Capasso had claimed to Monroe that he felt further experiments might be useful, a line that other MTSs remember Capasso taking in response to questions about Schön's work, Capasso did not, in fact, try to resolve this allegation by asking for more experiments, but referred to the data in hand. This is hard to reconcile with a claim disseminated by Cherry Murray, Capasso's manager at the time, after Schön's disgrace, that it did not "occur to his managers that deliberate fraud might have taken place."[17] (Murray continued to stand by these remarks when I asked her for comment.) Yet Monroe had been explicit about the possibility of intentionality, and Capasso had understood and reacted seriously, while John Rogers, Schön's immediate manager, ran his own calculations testing the allegation, and agreed there was a troubling problem. When Capasso wanted to raise the concern to the higher level of Murray, Rogers wanted to talk to Schön first, and Capasso agreed.

As the managers began to respond to the allegation, Schön's work continued to be promoted at the top of Lucent. On November 25, five days after the allegation, the company's chief executive officer, Henry Schacht, put his signature to a statement that gave Hendrik Schön and Zhenan Bao's work on the SAMFET as an example of Lucent's strong technological base. The statement and a photograph of the two researchers were published in a company report issued the following month, and a description of the research was added to a summary description of Bell Labs in the annual report filed at the U.S. Securities and Exchange Commission as well.[18] Schön's work also continued to be the subject of external scientific debate. On November 23 Schön received a note from *Nature* explaining that Paul Solomon, an expert on silicon devices at IBM Thomas J. Watson Research Center in Yorktown Heights, New York, had submitted a letter for publication that criticized the SAMFET paper. Solomon and colleagues were among those who had made inquiries to Monroe in October. Their letter summarized many of the serious technical questions that had arisen but did not discuss the data from *Science* that Monroe had questioned. The letter ended with the insistence that the technical problems were "profound," and an implicit accusation that Schön and Bao had been obscure about them. "It is a pity the authors did not come up with a plausible explanation for this, or at least comment on the lack of such explanation," the letter said. *Nature*'s associate editor Rosalind Cotter seemed to agree. She let Schön know that the journal was likely to print the letter and asked him for what she called a "clarification or correction" of his work to be printed in response.

Unlike Monroe's concern, which was internal and confidential, the IBM comment was expected to become public, and managers oversaw the Bell Labs response even before they confronted Schön with the allegation of data manipulation. Schön and Bao wrote the response, Bao pushed hard to try to meet *Nature*'s deadline, Rogers went away on travel, and Capasso suggested revisions so the response could go back to *Nature*. At one point, Solomon tried to call Capasso, who was a friend of his, hoping to get some more background on the human side to Schön's work, but the two failed to make contact.

During this time, Schön was not acting at all like someone who understood that he was being accused of violating a profound taboo. He sent the IBM letter to Monroe and cheerfully asked for help responding. He also sent Monroe two further manuscripts about SAMFETs. Monroe was drawn to respond, saying crossly that he agreed with IBM and repeating a scientific criticism that he had made earlier, that Schön had not shown any data revealing how the current through his devices varied as the width of the channel between his electrodes changed. Such a cross-check would be able to confirm, or rule out, the possibility of the artifact of an additional "parasitic transistor" forming elsewhere in the channel than where Schön thought. Meanwhile, the fraud allegation stayed unaddressed through early December, when Schön traveled to Berlin to receive the Otto-Klung-Weberbank-Preis, an award of DM 50,000 endowed by the Otto-Klung-Stiftung at the Free University of Berlin and the Weberbank Society.[19] Horst Störmer, who had been regarded as something of a wunderkind in his own days as a young scientist, also went to Berlin, and was photographed handing Schön a check several times larger than a baking tray. The comparison with Störmer, who had won the Otto-Klung prize in 1985 and the 1998 Nobel Prize for Physics, was not lost on the university newspaper, which published a long report on Schön's work under the heading "Tipped for a Nobel Prize."[20] The German language paper quoted Störmer telling Schön in English, "welcome to the club."

By the next week, December 10, Schön and Bao had sent their response to the IBM comment back to *Nature,* but managers still hadn't asked Schön for a response to the fraud allegation. In some way I could not track down, word began to spread that there seemed to be some kind of inquiry under way. A contributing editor at *Physics Today,* Barbara Goss Levi, who had written positive reports about Schön's work, began looking into the fact that his work had not been replicated by other groups, and on December 13, Bertram Batlogg, Schön's coauthor in Zurich, who was not involved in the SAMFET work at all, received an inquiry asking if he had heard anything about an investigation into Schön at Bell Labs. Soon after, John Rogers broached the subject with Schön, discussing several explanations for how Schön might, in all innocence, have ended up producing the overly symmetrical bar chart. Schön seemed to understand what was needed, and within a couple of days, had a response ready: a fabricated dataset that was noisy enough to pass the chi-squared test that his published bar chart had failed. Schön also described a way of getting from the dataset to the published bar chart that had involved some the data being mistakenly hidden from view. When this mistake was corrected, Schön explained, the bar chart did not look so unrealistically symmetrical after all.

To begin with, Rogers was uncertain and made clear to Capasso that he felt uncomfortable pursuing the issue. But on December 19, Schön sent the supposed original data to Don Monroe, who examined it, agreed that it contained more realistic noise than the published graph, and sent an email back to the managers in which he said that although the way Schön had got from his raw data to his graph sounded silly, it was not serious misconduct. In reaching this conclusion, Monroe assumed that the dataset he was receiving from Schön was known to be genuine, and did not go back and double-check whether the mistake Schön described could really have led to the published graph. Later, managers and investigators, including Rogers and Monroe, revisited this conclusion, and found that Schön's "hidden" data had never been present in the first version of his plot. But at the time Monroe captioned his email "CLOSED" and said how relieved he felt that they could go back to discussing purely scientific questions, forgetting, for the moment, that his second thoughts were usually a better guide than his first.

In contrast to the hesitation with which the managers had approached Monroe's allegation, they acted on the "CLOSED" email right away. Rogers forwarded it to Simon, to Horst Störmer, who had apparently also been informed about the concerns, and to Bertram Batlogg, saying that Schön had provided a "superb" response to what he called the "criticisms" made by Monroe, and that Schön would send a letter to *Science* to clarify. Although Cherry Murray later claimed in print[21] that Schön was apologetic and was reprimanded, email I have seen from the time suggests that, if anything, Rogers felt apologetic, saying to others that he hoped Schön had come through the issue without too much trauma. Capasso asked Monroe to let anyone who had been aware of the possibility of wrong doing to know that he had withdrawn his allegation, which Monroe did. At the same time, Schön resisted sending any explanation to *Science.* Prior to the closure of the allegation, Capasso had been worried about whether *Science* would request Schön's original data if the issue was brought to their attention—something that could have been a major problem for Bell Labs if, as Cherry Murray later wrote, Schon's failure to keep good records of his raw data had come to the attention of his managers.[22] But as far as Schön was concerned, no one had complained to *Science,* so there was no need to act. Rogers did not force the issue, and neither did Capasso. In the end, the admission that an analysis in a recent, high-profile paper was flawed was not shared with the rest of the scientific community.

When I asked to arrange an interview with Capasso and Rogers about what happened here, both declined. Capasso sent me an email saying that he had appointed a panel of "hard-nosed" scientists to investigate as soon as "the first serious and substantiated allegations" of fraud surfaced, a clear reference to a different set of external allegations brought to his attention six months later. This statement appeared to deny the existence of this internal investigation, even though Don Monroe's allegation was taken seriously at the time, so I replied to point this out. Capasso then telephoned me, suggested that he would be misrepresented if his blanket attempt to characterize the situation was taken to be a comment on Don Monroe, and went on to say that scientists are good at detecting genuine technical errors but not at detecting fraud, because their system is based on trust. "Yes, we were all fooled," he said. He added that for the sake of "peace of mind" he did not want to talk further and hung up. I received further information about Capasso's view of internal investigations second-hand in 2006, in the form of a note from a physicist who did not know anything about Capasso's involvement in the Schön case but who knew that I was interested in misconduct and wanted to let me know that a senior scientist, who turned out to be Capasso, had been vocal at a recent workshop, criticizing another institution that was in the news over fraud allegations by saying that the right thing to do was to set up a panel of external investigators, rather than trying to handle the matter internally. The physicist used quotes to indicate the reason Capasso gave: internal investigations "can never be trusted to get to the bottom of the issues involved."

Capasso did not tell other scientists at the workshop about his own experience of being involved with an internal investigation, but the implication of his remarks would be that internal investigators might be too close to an accused person to proceed in a sufficiently "hard-nosed" way. At the time of the internal investigation into Schön at Bell Labs, both Capasso and Rogers were closely involved in promoting his work. Although neither had coauthored with Schön, Rogers had put his signature to an oath for the U.S. Patent Office certifying that he coinvented the SAMFET. The existence of this patent application was disclosed by mistake, following Schön's disgrace, when the U.S. Patent Office published an abandoned application[23] on which Rogers, Bao, and Schön were listed as coinventors. In April 2003, Lucent Technologies tried to get the application withdrawn. The company sent a letter to the U.S. Patent Office titled, "Improper Publication of Abandoned Application" and went on to explain that according to U.S. patent law, "applications that have been abandoned should NOT be published." The U.S. Patent Office file of correspondence for another Schön application that was mistakenly published after abandonment,[24] on which Bao and two other scientists were coinventors but Rogers wasn't, contained no attempt by Lucent to get that mistake reversed.

Although Capasso was not on the patent, he had overseen Schön and Bao's response to the IBM letter to *Nature*, in the same period in which Monroe was bringing forward his complaint. This response turned out to be very effective. *Nature* sent the exchange to a reviewer of the SAMFET paper, who said that IBM was raising questions similar to those raised during the review process, and tried to defend his or her thinking in having recommended the paper for publication. Apparently at the request of the editor, he then reviewed the exchange not so much on the basis of IBM's letter, with which he said he agreed, but on the basis of whether the response from Bell Labs was worth publishing as a kind of "correction" to the original paper. In their response, Schön and Bao said that they did not claim to have made a conventional field-effect transistor, but to have unprecedented results that could only be understood through further research. They went on to mention additional experiments that Schön had supposedly done to rule out the possibility that his results were an artifact, and they ended by expressing their intention to publish a more detailed paper in another journal in the future. The reviewer advised the editor that nothing would be gained by publishing the "correction" and *Nature* got back to IBM to say that the entire exchange would not be published and to suggest that Solomon pursue his concerns in a more specialist scientific journal.

"IS THIS CREDIBLE?"

Following the closure of the fraud allegation and the rejection of the IBM letter by *Nature*, discussions about Schön's work continued at a more muted level. In January 2002, Chandra Varma, the theoretical physicist who had been trying to understand Schön's results, invited several Bell Labs MTS to a seminar about the SAMFET. Rogers and other managers did not attend, but the meeting was an attempt to bring together people in the Physical Research Lab with different views. Schön came and showed data to a group that included Nikolai Zhitenev, David Muller, and Rafi Kleiman. In this presentation, Schön changed the size of his gate insulator in the SAMFET from the thirty nanometers he had written in his papers to four or five nanometers, bringing it much closer to the size of the channel it was meant to modulate. This was consistent with earlier speculation by David Muller that perhaps the insulating layer in the device was thinner than Schön had previously allowed. Schön also showed data answering a question asked by Don Monroe in November as to whether the current through the devices depended on the width of his transistor channel; it did, ruling out the possibility of the artifact that had worried Monroe. These changes might have been expected to reassure the critics, but Schon was met on this occasion with a surprising amount of silence. Muller said that he left the meeting feeling that it was a mistake to engage too closely with Schön, who seemed to change his description of his devices in response to feedback, creating a moving target for discussion. "I said, 'wait for it to get into print before you criticize it'," Muller said.

That same month, Schön also traveled to the IBM Thomas J. Watson Research Center, where Solomon had invited him to come and discuss issues raised by the rejected letter to *Nature.* There, Schön described transistor switching behavior that was so finely tuned, sensitive to just a few hundredths of a volt on the gate electrode,[25] that Solomon, who had previously criticized Schön's work only on scientific grounds, began to feel that the work might well also be fraudulent. Fraud remained a topic of discussion at Bell Labs too, at least in jest. When Doug Natelson, a former postdoc of Bob Willett's who had become a professor at Rice University in Houston, visited in January, he joined a lunch table in the Murray Hill cafeteria and said lightheartedly about the SAMFET, "Is this bullshit, fake, or what?" raising a laugh. Natelson said that he and Willett talked about the interpretation of the SAMFET as scientifically very suspect. Yet there were few direct attacks on Schön, in part because of his amenable personality. "No one had it in for Hendrik. He didn't have enemies. It's probably one of the reasons it went so far. If he had been confrontational someone would have got mad and had the incentive to go after him," said Kleiman.

Instead, something more insidious happened. Unease about Schön's work fueled tension between people who doubted the work and Bell Labs managers who remained supportive of it. In February 2002, the tension exploded into a voices-raised argument between Bob Willett and Federico Capasso in the corridor of one of the research buildings. Willett had cowon the American Physical Society's Oliver E. Buckley Condensed Matter Prize, a $10,000 award for outstanding contributions to the advancement of condensed matter physics, and Bell Labs was organizing a symposium in his honor. Willett was upset to see Schön scheduled as one of the speakers and confronted Capasso, saying that he would prefer not to have Schön speaking at the prize event. Willett felt, based on his own experience, that there was no way to evaporate gold electrodes as neatly as Schön claimed to have done in the SAMFETs. Capasso's response was that

Schön was doing some of the most high-profile work in the lab, and that he would speak. When the symposium finally took place, on February 22, enough people were skeptical about the validity of Schön's work that Kleiman said he saw heads shaking in the audience as Schön showed a vast miscellany of data from various SAMFET and other devices. The theoretical physicist Bob Laughlin had been invited and spoke up in the question time to ask Schön whether his work was ever going to be replicated, although he stopped short of alleging fraud. Don Monroe was there too and felt so provoked both by the seminar and by a glowing profile of Schön that came out in the *Star Ledger,* the local New Jersey paper, on the following Sunday, that he sent Rogers a note saying that he could not imagine how anyone could possibly believe Schön had really done all the work he claimed. Monroe added that someone had shown him Schön's plot of the relationship between the current through his devices and the width of the channel (the data that he had suggested Schön could take to rule out the possibility of an artifact in the form of a parasitic transistor) and that it was so much more systematic than anything he had seen before that he was as suspicious as ever, he said. He added that managers had more than enough reason to set up an adversarial investigation into Schön, although it was probably too late to salvage Bell Labs' reputation "if things are as bad as they seem." Sensing that managers might not want to investigate one of their own, Monroe suggested that they bring in third-party investigators and run an investigation at arm's length.

This later became the route that Bell Labs' managers followed, but in February 2002 Rogers appeared to brush it off. His response to Monroe was that he did not think Schön's productivity was, in itself, a problem. He had discussed Schön with others, including Capasso, Batlogg, and Störmer, he said, and they did not think so either. He said that Störmer's advice had been "let Hendrik run." Störmer's philosophy was that young scientists at Bell Labs benefit from being given long stretches of time to focus on their research, free of management bureaucracy.[26] Writing as a relatively junior manager, Rogers made clear that he had understood the culture at Bell Labs prohibited him from interfering with an MTS's research, and concluded, "while I am not averse to an investigation, any direct involvement on my part would destroy my professional relationship with Hendrik."

Although Rogers was on the front line of complaints about Schön's work, Capasso was also aware of the growing problems. For example, Capasso welcomed Schön's collaboration with scientists outside Bell Labs as something that might help to shore up the MTS's flagging credibility. On the inside, Schön's collaborations were in flux, as several MTSs who had coauthored with him had not been able to follow up with him or share samples, and did not work with him again. Schön co-opted new people, including Chandra Varma, who was due to be a coauthor on one unpublished manuscript on cuprates, and Girsh Blumberg, an MTS with a background in high-temperature superconductors, who was due to be a coauthor on another. Both made suggestions for experiments Schön might try, and both were delighted by the data Schön brought back. Blumberg said he felt convinced by Schön's cuprates data because some features in the data seemed extraordinarily subtle, and yet were understood only by a small number of experts in the world, not including Schön. This idea soon spread around the lab as a response to the apparently growing suspicion that Schön could be faking data.[27] But as fast as it spread, it was undermined, as other scientists noticed subtle features that appeared to be peculiarly out of place. In 2001, for example, Laura Greene, a physicist at the University of Illinois at Urbana-Champaign, noticed a subtle kink in some of Schön's graphs on superconductivity in cuprates of the type that would be expected for a material with an organization of oxygen atoms different from the arrangement in the material Schön claimed to have studied.[28] It later turned out that Schon had fabricated at least some of the data in the paper.[29] Although Greene did not realize the data were fake, she did think that there had been a serious mix-up, and she brought the kink to Batlogg's attention when she visited him in Zurich in November 2001. Greene might have been well placed to bring up the possibility of misconduct too; in 2001, she had found irregularities in the data of a student and turned to institutional officials for help, triggering a process that culminated in a finding of misconduct against the student. But, she said, she did not, in part because the situation with her own student was ongoing and confidential, and in part because she hadn't quite confronted the idea that what had happened in her own lab was misconduct.

But even with the taboo on fraud stifling discussion, the scientific objections kept coming. On February 21, one of Schön's French coauthors, Catherine Deville Cavellin, came to visit Murray Hill. As was commonly done when outside scientists visited, Schön arranged a seminar and made a series of appointments for Deville Cavellin with other MTSs. Deville Cavellin said she was surprised to find that some of the people she met did not seem to appreciate Schön, and she was disappointed that Schön did not seem to have the copper oxide samples she had sent him. In her seminar, she showed images of the samples taken in Paris using an electron microscope, which raised the interest of David Muller, who had been trying to get hold of samples from Schön to image with his own instrument. From the images, Muller could see that the surface of the copper oxide seemed to have reacted with the air, creating a dead layer. No matter how good Schön's deposition of aluminum oxide, the gate voltage would still have to work through the dead layer. "You can't gate this stuff," Muller thought. After the talk, Muller asked Schön what he had done to treat the surface of the material before depositing the layer of insulator and making the field-effect transistor. Schön said that he had not done anything. On the one hand, colleagues such as Blumberg (who had provided Schön with a different sample from those shown in the talk by Deville Cavellin) seemed totally convinced by the plausibility of Schön's work. On the other, it did not seem physically possible to get some of his data from the samples that Deville Cavellin had provided. Muller said that he walked back from the talk to his office feeling sick.

As concern mounted, managers made some effort to respond. Rogers and Capasso began the process of recruiting a postdoc who would work with Schön to replicate his findings. Getting a postdoc would usually be considered a perk, but one former MTS who said he had participated in a review of research for Cherry Murray that resulted in the idea being put forward told me that the idea was not to reward Schön so much as to get in extra manpower to help clarify his groundbreaking claims. Schön was also beginning to realize that managers were not entirely happy with his claims, at least judging by his public presentations. His talks and papers frequently ended with a section titled "Acknowledgments," in which he credited close collaborators and name-dropped references to senior scientists who had spoken with him. The suggestion that people listed in the "Acknowledgments" might agree with his results was made more explicit in one email I saw; upon attaching a manuscript in which Horst Störmer was acknowledged, Schön noted that Störmer had encouraged him to write up the enclosed data. But by early 2002, Schön had added a separate section to the end of public presentations titled "Discussions" and listed the names of John Rogers and Federico Capasso there, rather than under the "Acknowledgments." He had realized, had perhaps even been warned, that he should not imply that people overseeing him necessarily endorsed his controversial results. Over the following months, the hold that managers had placed on the single electron transistor paper remained in force, but a number of Schön manuscripts on other topics continued to ship out.

11

GAME OVER

A couple of weeks before the annual March meeting of the American Physical Society in 2002, Hendrik Schön gave what would turn out to be his final attempt to demonstrate the sputtering of aluminum oxide onto organic crystals in Konstanz. He did so at the request of Bertram Batlogg, who had made a sensitively worded appeal to try to convince Schön to share his technical tricks with others. Batlogg did not go to the demo himself, traveling instead to his home county of Vorarlberg in Austria to give a scientific talk, but according to researchers from Batlogg's laboratory in Zurich who attended, it was a miserable failure. Schön did not make any working devices, and he failed to impress the others with his lab technique as well. Jun Takeya, who was spending a year in Batlogg's lab as a visiting researcher and had obligations to his home institute, CRIEPI in Japan, to get at least some useful results during his time away, made clear to Batlogg and others that he was giving up on sputtering, and he convinced Claudia Goldmann, Batlogg's graduate student, to do the same. Like researchers in Delft, Takeya had begun making field-effect transistors from organic crystals in a different way: placing organic crystals gently on top of a gate insulator and gate electrode, rather than trying to deposit the insulator and electrode on top of the delicate crystals.

As members of his own laboratory lost confidence in Schön, Batlogg stayed as upbeat as possible, telling one researcher in February 2002 that other researchers might replicate Schön's claims even if his own laboratory could not. But colleagues who knew him well told me that he seemed worried and perplexed. On the weekend of March 16 he arrived in Indianapolis for a workshop on organic materials scheduled to coincide with the American Physical Society March meeting. He met later with Christian Kloc and asked whether it was possible that Hendrik Schön had never made field-effect transistors. Kloc said no, it was not possible. Kloc recalled going into the lab shortly before Christmas 1999 and finding Schön exhausted, having been up all night obtaining the first quantum Hall effect data. There was no way that Schön had not made a field-effect device, Kloc said. Something else lay behind the problems they were having.

Schön's collaborators concluded that all they could do was keep trying to persuade Schön to publish his sputtering tricks. There still seemed a reasonable chance of replication elsewhere. Although Schön would not share his samples, Christian Kloc had sent his organic crystal samples out to as many as two dozen other research groups. None had come back with definitive reports of replication, but people were always delighted to receive the crystals in the mail and their immediate reaction was often to send Kloc positive comments. On Sunday evening, Batlogg felt relaxed enough to leave the conference center for a concert by Art Garfunkel, who was playing in Indianapolis that night. Batlogg was a fan of the folk-rock and for a time even hosted a clip from the 1967 movie *The Graduate* (which had a soundtrack featuring Simon and Garfunkel) on his lab web site. The clip showed the scene in which an exuberant family friend advised the main character of the movie, Benjamin Braddock, on how to make his fortune by saying "just one word, Plastics." Not having known Garfunkel would be in town, Batlogg had not gotten tickets in advance, but to his surprise, he got in.

Within a day, the fortunes of the collaboration seemed to be turning around. Back at the meeting, Batlogg bumped into one of his former postdocs, Art Ramirez, who had been trying to replicate Schön's work at Los Alamos National Laboratory in New Mexico. Ramirez said that the week before the APS meeting, his postdoc Vladimir Butko had seen sharp drops in resistance in a field-effect transistor made from an organic crystal sample sputtered with aluminum oxide. When Butko showed the data as part of his talk, the audience started fizzing with speculation about the possibility that this was a replication of Schön's work on superconductivity, Ramirez said. Batlogg was overjoyed to hear that one of Schön's experiments might have been built on, and he went out for a celebratory meal with Ramirez, Butko, Kloc, and at least three other Bell Labs MTSs that night. "We kind of got swept up in it," Ramirez explained. Batlogg also caught sight of Barbara Goss Levi, a senior editor with *Physics Today*, who had been making inquiries about concerns over Schön's work the previous fall. He wanted to call her over to let her know that, finally, the work seemed to have been replicated, Kloc remembered.

But the replication rumor never developed into the sure result that everyone had leaped on. Back in Los Alamos after the meeting, Ramirez and Butko tried carefully to replicate their original signal but found that they couldn't. They assumed it had to have been an artifact of some kind. Some months after that, they did see some reproducible drops in resistance in a pentacene crystal, and again wondered whether they had seen superconductivity. But when Ramirez demonstrated this effect to a colleague, Joe Orenstein, the subject came up as to whether the atmosphere in the cryostat might somehow be producing an electrical signal, an artifact that could masquerade as superconductivity. Ramirez said he didn't think so. He had a highly conducting sample in the experiment and pressed the button to pump out atmosphere, trying to show Orenstein that doing this made no difference. Instead, as he pumped, the apparent resistance of the sample began to rise. Molecules of helium, which the group had been using in liquid form to cool down their sample, were getting into the atmosphere and becoming charged, creating an arc of electrical current inside the cryostat. This occurred at the temperature at which the researchers had seen the leap in conductivity. As the molecules were pumped out, the signal went away. Ramirez subsequently found that he could get the effect to appear in the experiment even when there was no pentacene sample included; it was an artifact. Although the signal that had prompted a replication rumor at the March 2002 meeting was a separate incident and was never fully explained, the discovery of a similar artifact showed how common it was for artifacts to masquerade as signals in superconductivity experiments, and the group chalked it up to experience.[1]

With a background in superconductivity, Bertram Batlogg knew only too well that apparent observations of the phenomenon often turned out to be artifacts, and he later got to hear that the Los Alamos signal wasn't robust. But when I asked him, in 2006, what he remembered from the March 2002 APS meeting, his instinctive response was to remember, "Art reproduced our results." He looked delighted, and I could not help saying "Great!" Over the next few seconds, his expression changed, as he added that it had not turned out to be genuine. "Anyway," he concluded, with a dour look. "That's how I remember that APS meeting."

LEARNING SCIENCE FROM THE *NEW YORK TIMES*

The rumor that Art Ramirez might have replicated Schön's work spread around the APS meeting like wildfire. It reached, among others, Lydia Sohn, a professor of physics at Princeton University. Sohn was a gregarious person who described herself to me as "always talking on the phone" to scientific colleagues and friends. "My friends and I were the kind of people who didn't sit on the fence, said what we thought. We said when we were wrong too," she said. Before arriving at Princeton as an assistant professor, Sohn had held two postdoc positions: one at Bell Labs and one at Delft University of Technology in the Netherlands, and she had a PhD from Harvard University. She had a good publication record and was seen as a reliable critic, but she was not a superstar. She was often asked to review for *Nature*, for example, but she had never published a research paper either there or in *Science*. She had friends from each place where she had worked, but she had found it less easy to fit into the physics department at Princeton, where she found the more traditional physics faculty unenthused by the kind of careful applied nanotechnology and biophysics in which her lab specialized. For this, and for other reasons, Sohn told me, she had been denied tenure at her Princeton promotion hearing. The tenure denial meant that if she did not find a professorship at another university, she might have to leave academic science, and she definitely had less of an interest in the scientific status quo than a more senior researcher might have done.

Sohn first developed concerns about Hendrik Schön's work by talking with Leo Kouwenhoven, an expert in nanoelectronics she knew from her postdoc days in Delft. When Schön's first papers on the SAMFETs came out, Kouwenhoven asked Sohn what she thought. When Sohn gave a noncommittal answer, Kouwenhoven said something like "Come on, Lydia, it's fake data." Kouwenhoven pointed out the conductance data, the beautiful symmetrical bar chart that Schön had published in *Science.* (Neither Kouwenhoven nor Sohn knew that this data had been the subject of an internal allegation and investigation.) Sohn looked at the data and was shocked. She later described herself to me as having had a moral, sheltered upbringing. She felt astounded by the idea that someone might fake data to get ahead in science. Not long after, she saw Hendrik Schön in person when he came to Princeton to give a talk. Sohn attended, but found it hard to sit and watch as Schön paged through graph after graph of beautiful data from a wide variety of different nano-devices. When he showed some conductance data taken on his single molecule devices, Sohn said she spoke up. "I said, 'excuse me, how many devices have you actually measured?'" she remembered. Schön gave some answer. Feeling that the audience didn't sympathize, Sohn fell silent. But she was backed up by Giacinto Scoles, an older and sometimes belligerent professor of chemistry, who spoke up to say that with the technology of the time, it was not possible to get self-assembling molecules to arrange themselves in the orderly way Schön had described, with one end sticking to one electrode and one to the other, as at least some of the molecules would end up lying flat, both ends sticking to the same electrode. Scoles added, according to another attendee, that he felt he had learned more about Schön's work from the *New York Times* than from his scientific publications. Schön apparently smiled, as if he did not realize that this was an insulting thing for one scientist to say to another. Lydia Sohn said that she left the room feeling that the Princeton audience was more impressed by Schön than by the criticisms of him, especially when another faculty member came up to her on the way out and asked how many papers she had published in *Nature* or *Science* recently.

Along with Kouwenhoven, Sohn knew that other researchers in nanotechnology, including Paul McEuen, a professor at Cornell University in Ithaca, New York, were also worried about the possibility that Schön was fabricating data. And she understood through her contacts at Bell Labs, in particular her former postdoc supervisor, Bob Willett, something about the profile of Schön as a productive postdoc and whiz kid who had published an extraordinarily large amount of breakthrough work in several different fields. By November 2001 it was clear to everyone who looked into it that Schön was unusually productive. When Scoles was asked to provide a recommendation letter for Schön to be offered a chair at Princeton, he looked through Schön's publication list and then not only refused, but commented that Schön should be investigated for misconduct (Scoles later tempered his view). By the time of the APS meeting, Lydia Sohn was struck by the dramatic contrast between the ways that different people were reacting to Schön. She spent one evening in a Howard Johnson hotel bar with Bell Labs colleagues, who toasted Art Ramirez and the possibility that some of Schön's results had finally been replicated. Another evening, she went out for dinner with nanotechnology colleagues who were very disturbed by the same news. Paul McEuen in particular reacted with alarm, making the point that because Schön had published in so many fields, he might well have guessed some science correctly, and so have some work replicated even though he was faking. Although it is often assumed that fraudulent science will fade away when others fail to replicate it, skeptics of Schön's work were being confronted with a different scenario, in which wishful Bell Labs supporters looked as if they were jumping to the wrong conclusion about all the work from a hint of replication on one part of it. By the time Sohn left the APS meeting, she had talked to Bob Willett, her former supervisor. She asked Willett to keep her up to date about whatever happened next with Schön's work at Bell Labs.

"ALL INVERTER CURVES LOOK ALIKE"

Days later at Murray Hill, Julia Hsu and Lynn Loo stumbled across the evidence that spelled the beginning of the end for Schön. Both women were affiliated with John Rogers's Department of Nanotechnology Research. Hsu was a member of technical staff, and Loo was a postdoctoral MTS with Rogers as her designated supervisor. Neither was trying to blow the whistle on Schön. Instead, consistent with the need to save Lucent money, the two women had shared a hotel room at the APS meeting in Indianapolis and started talking about Loo and Rogers's work with soft lithography, a method for printing circuits onto plastic and other soft materials. Hsu asked Loo whether soft lithography could be used to make softer and gentler contacts with organic molecules, such as those Hendrik Schön was working with. On hearing that Schön wasn't interested in trying this out, she offered to work with Loo on the idea. After returning to Bell Labs, Loo and Hsu ran a series of experiments, and by April 19 they were ready to apply for a patent on the idea. In order to write the patent application, they needed to define how novel their work was relative to what was already known, so Hsu printed out a sheaf of relevant papers and grabbed them off the printer on the way to the Bell Labs patent office. The sheaf included Hendrik Schön's *Nature* and *Science* papers on molecular transistors. Hsu and Loo had just sat down with the patent attorney, John McCabe, when McCabe got a phone call, which he took, and they started leafing through the papers. Lynn Loo declined comment on what happened next, but a story that she has told that was relayed to me second-hand is that she was sleepy, nodding off, things going in-and-out of focus, when she noticed the duplicated data.

Schön's *Nature* and *Science* papers on the molecular transistors contained a couple of duplications, but the ones Lynn Loo noticed were the output of the inverters, the logical circuits that Schön had claimed to construct from his molecular devices. The outputs were identical, even down to the noise, the tiny wiggles caused by imperfections in the measuring process. As the *Nature* paper reported on devices made from a layer of thousands of electrically active molecules, while the *Science* paper reported on devices made from a diluted layer containing one or two electrically active molecules, the data should not have been very similar, let alone identical. Loo showed the duplication to Julia Hsu, who agreed that the curves looked surprisingly similar.

When she got out of the patent office, Julia Hsu took the two papers over to Mark Lee, an MTS with an expertise in circuits who was also her husband. Because Loo and Hsu had been talking about contacts, Hsu thought that maybe the similarity between the two curves revealed that Schön's transistor effect was an artifact caused by his contacts—which were the same in both devices—and nothing to do with the organic molecules, which differed. Unlike Loo, Hsu hadn't noticed that the noise between the two curves she was identical, but was interested in the scientific question of why the overall form of the data was similar. Hsu asked Lee what he though, and then went away, leaving the two papers with him.

Meanwhile, about an hour after finding the duplication in the patent office, Lynn Loo was working in the clean room with Bob Willett, both of them wearing masks, overalls, and caps to protect the devices they were working with from contamination such as pieces of falling skin or hair. As they worked, Loo told Willett that Schön had shown identical data, down to the noise, for two different papers and types of device. Willett asked Loo if she had told John Rogers, explaining that this kind of thing should be referred to management straight away. Loo said no, she was afraid. Willett said that if she didn't tell, he would. There was a phone on the wall of the clean room. Willett walked over to the phone and used the phone to call Rogers. Rogers did not answer, so Willett left a voicemail, telling Rogers that Schön had duplicated data between two different scientific papers. Later, after getting out of the clean room, Willett went and found John Rogers in person. He told Rogers not only about the duplication, but about what he thought was its significance. Schön had published dozens of papers in quick succession. If he was duplicating data, his work would be full of duplications. Reasoning in this way from a single example might seem like a large leap, but Willett had a qualification that was unusual among Bell Labs MTS. Before becoming a physicist, he had earned a medical degree. He had a good feeling for a wide range of possible human behavior.

As Willett was informing John Rogers, Mark Lee was looking at the curves too. Like Loo, Lee saw that they were not only similar, as Hsu had said, but identical. Because he knew it was impossible for two circuits to give identical outputs, Lee concluded that Schön had made some kind of clerical error and had sent the wrong data to one of the two journals. Before doing anything, he decided he ought to find out whether this was common knowledge. One person who would be likely to know was Bob Willett, as Willett was known to have a skeptical viewpoint on Schön's work. On Monday April 22, Lee came in to work and asked Willett whether the duplication was common knowledge. Willett said no. Mark Lee understood from this that Hendrik Schön did not know about it yet and that the right thing to do was to go and tell him, as he would tell any MTS who seemed to have made an error in a paper. (Lee and Willett's instinctive responses about who to tell about the problem differed, perhaps because Lee was making an assumption of honest error, while Willett was suspicious.) Some time that morning, Lee went into Schön's lab in Building One carrying copies of the *Nature* and *Science* SAMFET papers. He found Schön. He told Schon that the same inverter curves appeared in two different papers on the SAMFETs. Schön said no, the curves were not the same, only very similar. Lee disagreed. He showed Schön that the noise in both curves was identical, insisted that this could not be, and that Schön must have mistakenly sent the wrong file to one journal. Schön looked. Then he agreed with Lee that he had made a mistake and sent the wrong file to one journal. Schön seemed embarrassed, as if he really felt he had made a stupid mistake. Lee asked which curve was right and which wrong. Schön said he didn't know. Lee asked Schön if he would be so kind as to look this up and send Lee the correct data to replace the data that were wrong. Lee also said Schön would have to send an erratum to the journal to correct the error. Schön agreed. Lee left, and within a few hours, Schön sent Lee inverter data that he said should have appeared in *Science*; the figure in *Nature* was correct. The data in the erratum still looked very similar, so Lee began to think that Schön's result might be due to an artifact from the contacts, along the lines that Julia Hsu had originally asked him to consider. Some time later, Lee spoke to Bob Willett and told him that Schön was sending an erratum to *Science*.

Lee didn't follow what happened next, but Bob Willett and Lydia Sohn had been trading calls, partly about doing something social in connection with Willett's birthday, which had been over the weekend of April 20. Willett also knew that Sohn was trying to follow developments surrounding Schön's work. When he next called Sohn, she did not pick up, so he left a voicemail saying to watch out for an erratum on Schön's SAMFET papers. He did not say much more. He didn't need to. When Sohn picked up Willett's voicemail, she looked up the papers in *Nature* and *Science* and noticed the duplication. "It was obvious," she said. From her point of view, it also seemed unlikely to be an error. Sohn prepared all her papers carefully, and she couldn't imagine sending the wrong data to a journal, let alone mislabeling it to the extent Schön would have to have done, which involved changes to some of the numbers as well as to his caption. She created a slide in Microsoft PowerPoint that showed the duplication clearly, and sent it to colleagues, including Paul McEuen at Cornell, Charles Lieber and Charles Marcus, both professors at Harvard, and Leo Kouwenhoven in Delft. Marcus, who had previously taken a critical, but not an outright skeptical, view of Schön, said that as soon as he opened Sohn's file, he thought "game over. The file was definitely shocking." Sohn said the feedback she got was that someone had to speak to editors at the journals and let them know about the problem. Sohn was considered the best person to do this, in part, she said, because she was not planning a submission to either journal in the near future, and it was not known how editors were going to react. Sohn was also reviewing a paper for *Nature*, so she already had a line of communication open there.

Bob Willett's voicemail for Lydia Sohn was later discussed in hushed tones, as if it had been a mysterious leak by an anonymous whistleblower. The reality was more ambiguous. Willett was only passing on information that he believed was as good as public; Schön had told Lee he was sending an erratum to print in *Science.* John Rogers knew about the problem, and there was no reason for an MTS who trusted his colleagues and managers to do the right thing to doubt that the erratum had been sent. In practice, however, Schön had interpreted the problem as less urgent than the scientists on the outside did. On April 24, for example, Schön went so far as to submit a manuscript about the invention of a room temperature single-electron transistor (a paper that managers had previously asked him to hold until the SAMFET devices were better understood) to *Nature*, but by May 1 he still had not disclosed the error. While he hesitated, events on the outside began to overtake both him and Bell Labs managers. On May 2, Lydia Sohn sent a note to the editor she was reviewing for at *Nature,* pointing out the duplication between Schön's *Nature* and *Science* SAMFET papers, and expressing concern. Sohn's email arrived after closing time at *Nature's* head office in London, but on May 3 it was forwarded to Liesbeth Venema, the editor who had handled the SAMFET paper. Venema contacted Schön asking for an explanation.

On the same day that Lydia Sohn wrote to *Nature,* John Rogers, Schön's department director, also began to take a closer look at Schön's work. I do not know what triggered Rogers's action (Bell Labs apparently did not hear from *Nature* until May 3), but I interviewed one outside scientist who said he had been in the position of evaluating Schön and had called Rogers on May 2 with questions. The caller did not know about the duplication, but he was actively discussing Schön with a scientist who was in touch with Paul McEuen, who did, and a general impression of concern about Schön was spreading, even if the specific news of the duplication had not yet done so. Rogers did not disclose the duplication to the caller, but he strongly defended Schön, saying that Bell Labs knew the SAMFETs worked, although they did not know how they worked. However, that same day, Schön, under duress from managers,[2] finally sent his erratum through to *Science,* and Rogers began looking for further duplications in his work. Rogers logged a third SAMFET inverter curve in an unpublished chapter coauthored by Schön that looked possibly similar, although not identical, to the other two. He also examined the possibility of a duplication between two sets of transistor curves. A couple of days later, Rogers also called in Mark Lee and asked Lee to advise him on whether the data from Schön's SAMFET inverter could have come from the reported circuit. It later turned out that this was not the case; the curves were typical for a different kind of inverter, one presented by Schön in a paper about organic crystals published two years earlier, because Schön had used that paper as the source of the later data.[3] Rogers did not point out this additional data duplication to Lee but simply asked for advice as to whether the characteristics of the data in the later paper were physically possible.

John Rogers was therefore beginning to move forward cautiously with his own inquiries, but the next anyone learned on the outside was on May 8, when Liesbeth Venema at *Nature* got back to Lydia Sohn, Venema told Sohn that Schön had said that the data in *Nature* were correct, but that he had mistakenly sent data from the same inverter to *Science.* He provided her with a prepublication copy of an erratum intended for publication in *Science.* "All inverter curves look alike," wrote Venema to Sohn, using quotation marks to signify the explanation for the supposed mix-up that she had received from Schön (in the meantime, she had quietly rejected his April 24 submission). Venema provided Sohn with a copy of the *Science* erratum and asked whether she and her colleagues felt reassured. Sohn forwarded a copy of the erratum, and Schön's explanation, to Paul McEuen at Cornell. At this point, Paul McEuen got angry.

"I WAS LOOKING FOR PROBLEMS"

McEuen was known for his work building transistors from buckyballs, the soccer ball–shaped molecules made from sixty carbon atoms. He came from a conservative family in which he was the only vegetarian and the only political liberal, and he did not mind thinking of himself as the odd one out. He had found Schön's work impossible to ignore. To begin with, a couple of students in McEuen's lab followed up on Schön's claims by growing organic crystals, although McEuen had quickly gathered that the work was supposed to be nonreplicable and abandoned the effort without asking the students to try to make field-effect transistors. He had also tried to approach Bertram Batlogg at a scientific meeting in 2001, but had not been able to get Batlogg on his own, because, he said, too many other people were talking to him. "They were the golden people, you couldn't say anything," he said. Then, in November 2001, McEuen's lab held a journal club meeting to discuss the SAMFET papers, which had just appeared in print. One student spent time on a calculation trying to check some of the information in the paper. McEuen went on to give public talks in which he credited Schön's work on superconductivity in buckyballs as part of providing his audience with an overview of why scientists were excited about the materials. But he felt conflicted about giving attention to Schön's work. He suspected it was fraudulent because there was so much of it, because the SAMFET conductance data (the data that Monroe had questioned internally, and that Sohn and Kouwenhoven had been discussing) were too good to be true, and because of the replicability issues. The problem wasn't so much Schön, who he saw as "just a postdoc," but Bell Labs, a hugely respected institution, throwing its full authority strongly behind the shaky claims, in a way that provoked McEuen's political sensibilities. Unlike, for example, Stephen Forrest, a strong critic of Schön's at Princeton University who put aside thoughts of fraud and encouraged one of his students to analyze Schön's results critically, McEuen wanted to do the opposite, to tell his students not to work on anything related to the Bell Labs work because he believed it was fake. "My group was enthralled by this. I didn't know what to tell my students. I couldn't say, 'stop, because they're cheating'" he said.

The possibility that Schön's suspicious duplication was being dismissed by Bell Labs managers as a mere error was the final straw for McEuen. On May 8, the day that he received the erratum from Sohn, McEuen printed out a heap of Schön papers and began to take them everywhere, including when he went to get his sandwich at lunch and home at night. He said that to begin with he was not looking for duplications per se, only for evidence that he could point out in order to explain why Schön's work was wrong. This sudden fervor was later described to me by one colleague of McEuen's as nothing short of "brilliance." The way McEuen put it was simply, "I was looking for problems." On the evening of May 9, he talked to Lydia Sohn by telephone, the papers still in front of him. *Nature* had asked Sohn for an opinion on the Schön erratum, and Sohn thought that the new curves still looked very similar to the old ones. McEuen took on board this observation, staring at both the erratum and the duplicated curves until the noise wiggles in the curves became frozen in his mind's eye. Then, as he looked to the sheaf of papers in front of him, he caught sight of a page on which Schön had published the output from another inverter, and he saw the same wiggles again. This time, the data supposedly came from a circuit Schön claimed to have made from pentacene field-effect transistors in 2000, more than a year prior to publishing on the SAMFET. McEuen might even have been on the phone to Sohn in the moment that he saw it. "I said 'Lydia, I found another one'" he remembered. This was a third example, involving an entirely different material and, in McEuen's mind, conclusive proof of fraud.

But would it be enough to convince others? That night, May 9, McEuen stayed up all night searching for duplicated, inconsistently captioned data in Schön's work. Whenever he found something, he created a slide in Microsoft PowerPoint that could help make it clear what the problem was. He decided that he would like to settle the issue definitively by the coming weekend, when he had to fly to Maui, Hawaii, where he was scheduled to speak at the First International Conference on Nanoscale/Molecular Mechanics. During the night and next day, he and Lydia Sohn also exchanged slides and opinions with Lieber, Marcus, and Kouwenhoven. On May 10, McEuen walked into the office of Dan Ralph, a physicist in his department at Cornell, not having slept, and said, "'This stuff is fake'," according to Ralph. McEuen knew that Ralph was on a committee to choose the winner of the William L. McMillan Prize, an award of $2,500 given annually by the Department of Physics at the University of Illinois at Urbana-Champaign to a young condensed matter physicist within five years of gaining his or her PhD. Schön had narrowly lost out on the prize in 2001, on the basis that his work had not yet been replicated, but the committee had picked Schön to win in 2002, reasoning that even if half of his work turned out to be wrong, he was still ahead of the runner-up candidate. The choice had not yet been mailed to Schön or publicly announced, and when Ralph saw the evidence that McEuen had found, he got in touch with the rest of the committee and the prize letter to Schön was retrieved from a secretary's out-tray at the UIUC Department of Physics.

As word of the duplications spread, physicists around the world started searching for duplications. McEuen reduced his own list of concerns to six duplications in five papers (two of which had been the subject of the erratum). He left out concerns that weren't compelling, for fear of making his case look petty, but he included enough to show evidence for a pattern. He had in mind only the need to convince others that there was a problem in Schön's work, but he wasn't calling for any particular course of action to be taken. Having found the evidence, he decided what he thought was the right thing to do, and then he did it. On Friday May 10, McEuen wrote to and called Schön to allege that Schön had faked the data. He tried to reach Bill Brinkman, who until September 2001 had been the vice president for research at Bell Labs, and Federico Capasso, the vice president of the Physical Research Lab. He did not reach them. He did reach John Rogers by phone and made clear to him that he planned to inform the academic journals in which Schön had published, and he sent Rogers the evidence that he had collected. He also wrote to Schön's long-time coauthor, Bertram Batlogg, in Zurich about what he called "grave concerns." He attached some of the evidence, and asked that Zhenan Bao and Christian Kloc be informed. Finally, he wrote to editors at *Nature* and *Science*, and then he left for Hawaii.

"I HAVEN'T MANIPULATED DATA"

Paul McEuen's allegations, along with his intention to make the matter public, were referred up the management reporting line at Bell Labs from John Rogers to Federico Capasso and Cherry Murray. At this point, the managers decided to open an external investigation into Schön. This was not a new idea; it had been suggested by Don Monroe to Capasso in November 2001, and repeated by Monroe to Rogers in February 2002. But this time, the managers acted. Over the weekend, Capasso and Rogers met with Schön and explained that they would set up an investigation. Internally, the idea was spun positively; the investigation might help to clear Schön's name if he had acted with integrity, although both Capasso and Rogers had their doubts. Capasso began to ask why Bertram Batlogg, who was no longer at Bell Labs but had coauthored with Schön, had never seen any of Schön's remarkable results (these had not been seen by managers either). On Sunday May 12, Capasso telephoned Batlogg at home in Zurich to tell him about the investigation. He explained the situation to Batlogg by saying that allegations were mounting everywhere and that the lab had no choice. Batlogg told Capasso that the investigation should include a German-speaking member, who could help communicate with Schön. Capasso asked Batlogg if he would ask Schön the question that no one at Bell Labs had asked: whether or not Schön had manipulated data. Batlogg said yes. Late on May

12, Batlogg called Schön, and spoke to him for about 105 minutes, taking longhand notes on a letter-size notepad. They talked through the situation, and Schön filled in Batlogg regarding concerns over the SAMFET papers, on which Batlogg had not been a coauthor. Schön told Batlogg that he was being unfairly attacked, that managers had made him cancel his public talks, and that he was feeling overcome. Batlogg asked Schön whether he had manipulated data. Schön said no. Batlogg told Schön to cooperate with the investigation.

Over the next week, Bell Labs managers assembled a panel that included two German speakers: a department director from the Bell Labs site in Holmdel, New Jersey, Herwig Kogelnik, and the cowinner of the 2000 Nobel Prize for Physics, Herb Kroemer, of the University of California, Santa Barbara. They also included Supriyo Datta, a physicist at Purdue University who had been following the work on SAMFETs. As the chairman, they selected Malcolm Beasley, a distinguished superconductivity expert at Stanford who had coauthored a technical comment in *Science* that questioned samples grown by Schön's French collaborators. The managers also invited Don Monroe. This was an unusual step for an impartial panel, given that Monroe had previously made allegations against Schön, but while Monroe had made an allegation of intentional data manipulation that managers had not gotten to the bottom of, he also knew that he had withdrawn it. Monroe's appointment became a way for concerns shared by critics in the internal Bell Labs community to feel represented on the committee. With two critics of Schön and four outside scientists involved, managers could not be accused of appointing a committee that would be soft on Schön. They charged the panel with investigating him and his coauthors. Over the next two weeks, John Rogers worked on summarizing concerns known to managers to be handed over to the committee. By accessing the data behind the graph questioned by Don Monroe in November, Rogers also showed that the data Schön had used to respond at the time were not the same as the data he had published, showing that the issue had not been resolved correctly the previous fall.

As this was happening, journal editors were getting back to Paul McEuen. While editors at *Nature* were very concerned, manuscript editor Phil Szuromi at *Science* was slower on the uptake, initially putting forward the suggestions that McEuen might want to submit a technical comment on Schön's work, or consider getting a lawyer, both of which McEuen brushed off. Then, in a swift about-turn, Szuromi seemed to get it and asked McEuen whether he would be prepared to speak to the reporter Bob Service, who was already working on a long piece about Schön's work. The journal *Science* claims to operate a Chinese wall between its manuscript and news sides, but tips do pass back and forth.[4] McEuen was happy to talk and, by a stroke of chance, Service was in Hawaii covering the Nanoscale/Molecular Mechanics conference, so the two made contact. On May 20, Lucent's press office began to talk to reporters about an investigation, without naming names, and Service put an article about the duplicated data up on *Science*'s web site.[5] Three days later, news of Schön's identical curves was in the *New York Times*.[6]

THE PLASTIC FANTASTIC REVISITED

As Bell Labs managers made arrangements for an investigation, Schön's internal critics continued to follow up. Aware that his call to Sohn had triggered a remarkable chain of events, Bob Willett arranged to have a one-on-one meeting with Schön. As if drawing on his medical training in bedside manner, he explained carefully to Schön that his career was on the line, but added that Schön might be able to clear his name by producing samples to be independently measured. If there was any real science in Schön's work, it might be salvaged from the scandal surrounding the data irregularities if only Schön could produce a working sample, Willett figured.

Schön did not respond to the idea in the meeting, but on May 14, he arrived in the next-door lab of David Muller. He was carrying a sample, about two years after Muller had first asked him for one. In Schön's hand were several pieces of a glass slide, onto which he had deposited a molybdenum electrode and sputtered a layer of aluminum oxide. On one fragment was a layer of polythiophene deposited by Zhenan Bao to form a field-effect transistor. This was supposed to be the superconducting polythiophene, the "Plastic Fantastic." Schön told Muller that he did not have good notes about the sample, but that from memory the layer of aluminum oxide was 30 nanometers (billionths of a meter) thick and came from his "good" sputtering machine in Konstanz. Muller looked at the sample. The aluminum oxide shimmered with interference colors, like the rainbows on an oily puddle caused by light reflecting and refracting through a thin layer. Each rainbow revealed the refraction of light through around 100 nanometers. Counting by greens, Muller could make out three rainbows, suggesting the aluminum oxide was around 300 nanometers, ten times thicker than Schön had said. Muller took the sample out of Schön's hand into a perspex box, opened a new lab notebook and wrote down the date and the sketchy information. Then he went to Bruce Van Dover's lab with another MTS, Glen Wilk, to find out at what applied voltage the aluminum oxide broke down. With John Grazul, another MTS, he also took images of the field-effect transistor sample in an electron microscope, which revealed the dimensions of the aluminum oxide layer in the heart of the sample to be 250 nanometers, not 30 as Schön had said. Because the breakdown field of a material at a given voltage depends on its thickness—the thinner the layer, the larger the field—Schön's misstatement could have led Muller to infer that the breakdown field of the material was more than eight times its true value, had the information not been double-checked.

Over the summer of the investigation, Schön was not suspended from coming into work, but managers did put him on ice. They alerted the U.S. Patent Office to the questions that had been raised by Paul McEuen, wrote to the scientific journals in which Schön had published, and prevented him from making further claims in public. When one outside scientist, Tom Jackson, who was co-organizing a conference at which Schön had been going to speak, called Capasso to try to appeal the withdrawal of Schön, Capasso reacted negatively, shouting loud enough for Jackson, who had not yet realized the seriousness of the situation, to hold the phone away from his ear, Jackson said. As the extent of the fraud dawned, managers also became anxious about how Schön was feeling and what he might do. They offered him counseling, and rumors circulated that they were locking their office doors when they had to work late at night with Schön in the building. When the report came back at the end of the summer, managers released it to the public, and, on the same day, fired Schön. This made for a very public disgrace for Schön, cued his managers and coauthors to start retracting his papers, and conclusively demonstrated his separation from the rest of Bell Labs. It also clarified to scientists who had been waiting in suspense all summer that the remarkable string of breakthroughs had truly been a sham.

This resolution to the Schön case is sometimes attributed to the openness of Bell Labs managers. Both Federico Capasso and Cherry Murray expressed pride in the establishment of the investigation and the release of the final report in comments they made to me. In the scientific community today, such a level of transparency in the face of fraud remains rare, with the government agencies that sponsor physical sciences research frequently failing to ensure full and open investigations when allegations of fraud arise. At Bell Labs, however, it had taken an extraordinary confluence of events to bring the scale of Schön's fraud to light. Even before the first duplication had surfaced in the patent office at Murray Hill, at least half a dozen researchers, inside and outside Bell Labs, had kept suspicion high on the agenda, so that managers had been confronted with the possibility of data manipulation and had been warned—by Don Monroe—that only an external investigation might have a chance of saving Bell Labs' reputation. Then Paul McEuen and Lydia Sohn went public, laying out enough evidence for the problem to be appreciated, but without appearing to tell Bell Labs how to respond, so that the institution had room to appear proactive rather than defensive. The McMillan Prize committee, in the act of considering an application from Schön, had exerted subtle pressure by telling Federico Capasso that it would keep its decision on hold over the summer and that it expected to receive an update on the results of the investigation. Other scientists had made clear, both to the media, and to managers, that they were looking for clarification too. Then the investigation panel produced an unusually bruising report, which would have been hard to suppress from public release without risking accusations of a cover-up. Against this background, the handover of allegations to a commission that Cherry Murray promptly advertised as "a blue ribbon panel,"[7] and the release of the commission's findings to the public, was, more than anything, the smartest solution, possibly the only solution, to the problem that Schön had become. It was also a sure-fire way for Bell Labs to suggest that it was still capable of getting to the truth, even having produced a fraus on a spectacular scale. There was an additional factor driving the Schön case toward a public showdown of some kind: Hendrik Schön's single-minded commitment to keep trying to publish fabrications.

"I GUESS IN SEVERAL CASES IT WAS OBVIOUS"

By the time his duplications came to public attention, Hendrik Schön had been duplicating data for four years. To begin with, he duplicated the data that set the scene, introduced the field-effect device, before he told the story of each new breakthrough. Having found a stage set that others seemed to like, Schön kept using it, maintaining a consistent picture and feeling increasingly comfortable.

But by mid-2000, as he began to strike out on his own, Schön began to duplicate data that was central to his papers, running a greater risk of detection. When he was asked for additional laser data to satisfy internal review of his laser manuscript at Bell Labs, for example, he produced a duplicated spectrum that he substituted for another one later that year.[8] In two SAMFET papers, he duplicated not only transistor curves, but also an inverter circuit that appeared toward the end. The more unsure he felt, the more always he reverted to data that had worked out well for him on previous occasions. In December 2001, when responding to the IBM letter that questioned the invention of the SAMFETS, he insisted that he had measured identical results in more than ten different kinds of organic molecules, even though this wasn't something IBM had asked.

Facing questions about the SAMFETs, Schön also began urgent attempts to bring the devices back in line with theoretical expectations. He began surfing for ideas on "cond-mat," part of a web site hosted by Los Alamos National Laboratory that is used by physicists to share copies of papers prior to publication. On January 18, 2002, a theoretical paper[9] appeared that caught his interest. Eldon Emberly at NEC Research Institute in Princeton, New Jersey, and George Kirczenow at Simon Fraser University in British Columbia, Canada, had predicted the properties of an electronic device containing a single organic molecule sandwiched between two nickel electrodes. Nickel is ferromagnetic—it behaves like a little compass needle when a magnetic field is applied to it. Emberly and Kirczenow suggested that the resistance of a molecule placed between two nickel electrodes might change depending on whether the magnetization of the two electrodes was aligned, with their north poles pointing the same direction, or misaligned, with the poles in opposite directions. The device, called a spin valve, was a very simple magnetic switch that did not even need a gate electrode. Schön was hooked, and within about a week of the paper's appearance had sent off an email to Emberly and Kirczenow claiming to have measured the thing that they had predicted. Not being a chemist, he misread their paper and claimed to be using a different molecule from the one that was the subject of the prediction, but as soon as Emberly and Kirczenow got the message, they repeated their calculations for the molecule he described and found those were consistent. "I was amazed by the agreement," said Kirczenow. Although Schön had previously had no contact with the two theorists, his Bell Labs affiliation gave him enough credibility to impress, and by following theory, he could also be sure that the data he produced for this device would have a clear interpretation. When, in early February 2002, John Rogers asked what he was working on, he fired off two papers on the very same day, one to *Nature* and one *Science*, the second of which contained his agreement with the prediction that he had found on cond-mat two and a half weeks before.[10]

This was not the only example of Schön's shooting off in a new direction as he tried to bring his fabrications back into the fold of others' expectations. Following the first publication and presentation of the SAMFETs, other MTSs pointed out that because the channel in the SAMFETs was so small, the conduction of current probably happened by quantum tunneling of electrons from one electrode to another, rather than by conventional charge flow. Schön began to develop new data on tunneling, this time using molecules provided by the chemists Peter Bäuerle and Günther Götz at the University of Ulm in Germany, where he had been invited to speak in 2001. One new manuscript described his creation of a single electron transistor, a quantum device similar to the SAMFET but able to switch current one electron at a time. After November 2001, he was not allowed to submit this manuscript for publication, but managers did not say that he could not show it to other people, and so he kept sending it to colleagues in the hope of getting feedback, using tricks such as his trademark ellipsis of ". . ." to try and provoke friendly responses. By February 2002, his worry about others' doubting him had reached the point where, as he prepared to show some new data that he said might be pointing to the formation of a so-called Wigner crystal,[11] in which low-density charges on the surface of an organic crystal settle into a regular formation, he told a colleague that he might show the data without giving any interpretation at all, in order to see whether people in the audience would guess what effect he thought the data illustrated.

But, he realized, other scientists were beginning to tire of his approach. At the Materials Research Society spring meeting in San Francisco, where he accepted the Outstanding Young Investigator Award, a $5,000 prize, Schön hung out in a session where results were being presented on posters, just around the corner from a small crowd of people talking about his SAMFETs. There was a conversation between a Bell Labs colleague and Stephen Forrest that Schön could follow from a safe distance. Forrest was asked whether he believed the SAMFET result and replied that it was not a question of belief. The problem, Forrest said, was that Hendrik Schön seemed to be expecting the rest of the scientific community to do his heavy lifting for him, reproducing work that he hadn't done carefully to start with, and failing to provide a clear interpretation of his results, hoping that others might figure things out for him. After they had talked for a few minutes, Forrest left and came out into the main aisle. Schön tried to tell Forrest that Art Ramirez's group had almost seen superconductivity in an organic crystal, almost replicating his findings. Forrest recalled that this had seemed very unconvincing.

Feeling insecure about the reception he was getting from others, Schön also began to worry about his career plans. As well as being under consideration at Princeton, he was near to getting an offer from the Max Planck Institute in Stuttgart. Klaus von Klitzing had recommended that the Max Planck Society consider offering a research group directorship to Schön, who would have been the youngest ever to take up such a senior position. Feeling uncertain whether it was a good idea to leave the sheltered environment of Bell Labs, Schön asked Theo Siegrist, a materials scientist in the Physical Research Lab, for advice. Siegrist said that the Max Planck job was a once-in-a-lifetime opportunity and that Schön should take it. All of a sudden, people at Bell Labs seemed to be comfortable with the possibility of Schön's leaving. He was also losing his aura as a Konstanz-based sputtering expert, as work done by researchers from Batlogg's lab on the Konstanz sputtering machine had failed to achieve breakdown strengths any greater than about half that of those needed for his field-effect devices. With his own equipment under scrutiny, he could only save his own plausibility by admitting that things weren't working out, that he could not replicate some of his earlier results, telling a meeting of the American Chemical Society that he suddenly understood what others had been going through.[12] Finding himself at the end of a dead-end alley, he decided to turn around and follow the lead of others who had been following up on him. He claimed that he had started gating devices using insulators different from aluminum oxide, and he circulated at least one draft manuscript showing results on a device made with an insulating later of a plastic called tetratetracontane. His story about aluminum oxide had finally run its course.

But while Schön's method was timing out, he was able to keep going, changing his claims in line with the emerging consensus, until the day when Mark Lee came into the lab and pointed out the duplicated SAMFET inverter curves. Schön reluctantly agreed that yes, this had been a mistake, and then quickly shipped out to a journal the manuscript about the single-electron transistor device; on which managers had put a hold two months before. When Paul McEuen got in touch about further duplications on May 10, Schön responded that his different devices could easily have provided similar data, and sent McEuen examples of other scientists' inverter curves to try to prove it. Over the weekend, as Rogers and Capasso called him in to talk about setting up the investigation, Schön told them that unfortunately he did not have many records of his original data, or samples to show. Capasso became angry and sharply criticized him.

Still, Schön tried to keep going. He sent an email to Klitzing in Germany showing his new paper in *Science* about spintronics. On Sunday, he spoke with Batlogg, who told him to cooperate with the investigation. By Monday, May 13, he had reflected that maybe he should tell Klitzing about the investigation, because, as he explained, it was better that Klitzing heard about it from him than that he heard about it from others. He also told Klitzing that he had not manipulated data, that the investigation was going to show how little content the allegations had, and he ended by saying how difficult the situation was for him. Then everything went public.

During the investigation, Schön kept coming in to Murray Hill, and worked mostly on his computer as around him colleagues tried to replicate his results. They included Roswitha Zeis, a student of Ernst Bucher's, and Jochen Ulrich, the postdoc recruited by Horst Störmer. Schön explained to the others that his results were true and that he had really measured them; the unfortunate situation was only that he could not show it. Colleagues kept trying to address further issues with him. A researcher from Brookhaven National Laboratory, the site of the proposed Department of Energy Center for Functional Nanomaterials, on the proposal for which Schön and Zhenan Bao had been listed as participants, was invited by Theo Siegrist for a visit to talk about the SAMFETS. In a meeting with Schön and Bao, the outside researcher, Ben Ocko, asked what kind of silicon had been used to make the devices and how to obtain the organic molecules used. Schön and Bao had no useful information to offer. Christian Kloc was also under pressure and told Bao that he believed they would all be fired once the investigation was over; their manager, Elsa Reichmanis, came and told him and Bao that she would oppose any move to fire Schön's collaborators. Kloc kept putting his time into earnest replication efforts. By email in June, Schön told Bertram Batlogg in Zurich that people—"we," he said, although it was others doing the work—were still trying to replicate the results at Murray Hill, but with little success. He said he guessed Batlogg wasn't too happy with him at the moment. Then he met twice with the committee but it did not go well. They had started looking at some of the data he had fabricated using theoretical functions. He tried to tell them that he sometimes smoothed out his results, which sounded much better than making them up, but they did not seem to agree with that explanation. Anticipating a bad outcome, Schön began to warn collaborators, including Bertram Batlogg and Catherine Deville Cavellin, that the committee had not gone as well as he would have liked and, remembering the point about the poor recordkeeping, warned both that, unfortunately, his recordkeeping had let him down. He picked up on the discussion at Bell Labs, started by Capasso and others, about whether his collaborators had seen his results, and wrote to Batlogg that he knew Batlogg hadn't seen them all, but insisted they were real. He told Batlogg that he wasn't "in the money" any more and did not expect people to trust him again.

Once the investigation report was ready, Schön and others had some days to read it, under lockdown in a room at Bell Labs, before it was released publicly. Schön wrote his statement admitting to "mistakes" and said he hoped that he might be able to keep working for his results, if possible. Then John Rogers and Cherry Murray fired him. Schön told them, according to what different people later said, that he'd had a good time, or that it had been an honor working at Bell Labs, and left the building calmly, accompanied by two security guards, although, as one outside colleague later commented, one security guard would probably have been enough. He had a few days in Summit before leaving for Germany. In that time, he bumped into several people, including Saswato Das, the Bell Labs press officer, whom he called to from behind in the Summit train station and approached to say how sorry he was, Das later said. But in an email circulated to a colleague, Schön later sought to put the blame where he had directed his apology, writing that "I was never as good as Lucent public relations made out," nor as bad as the media were saying. The truth, he insisted, as plausibly as ever, was somewhere in between.

Back in Germany, Schön told friends and colleagues that he intended to keep working in science. He said that he would get another job and see his results reproduced in the future. When managers, coauthors, and editors began to retract his papers, he agreed withdraw the ones that were written up by the committee but pleaded with editors not to retract others. They did so anyway. Organizations also tried to distance themselves; the Materials Research Society removed the news of his Outstanding Young Investigators' Prize from their web site (only later putting it back on for archive purposes),[13] and the German papers ran negative news articles until he agreed to follow the lead of his collaborators and return his share of the science prizes he had won to the City of Braunschweig and others. Eventually, he listened to a former member of Ernst Bucher's lab, who explained that he could not go back to work in science and suggested a job at an engineering firm in Germany. Schön agreed. One year later, the University of Konstanz concluded that he did not commit misconduct in his PhD work but still took away his PhD.[14] The German Research Foundation reached a finding of misconduct and introduced a ban, so that Schön could not receive research funding for at least the next eight years.[15] Schön contracted an attorney and talked about trying to get his PhD back through the courts. He tried to stay in contact with former colleagues, but few people wrote back. Few people, he wrote to a colleague, wanted to know his side, or contacted him directly.

"I guess in several cases it was obvious," he wrote, in an email circulated to a colleague, that some of his results had been smoothed out. He added that although he had misrepresented data, he did not want to go into details of what exactly he had misrepresented, that he should not have done it, but could not make it undone. The problem, as he kept telling collaborators near the end, was that he did not have enough time. It was bad luck, and no one could be sure that things would not have turned out differently, with better luck, more care, and a little more time . . .

EPILOGUE

Once fraud had been exposed, it did not take long for researchers working on organic electronics to move on. Within a year, researchers at Delft University of Technology had reported on the observation of ambipolarity, the transport of negative and positive charges, in thin films of pentacene, the material in which Schön had claimed to measure the effect.[1] In 2005, two teams, one led by Henning Sirringhaus and colleagues at Cambridge University and the other by Alan Heeger and colleagues at the University of California, Santa Barbara, reported on light-emitting field-effect transistors in which positive and negative charges induced on the surface of an organic material combined and emitted light.[2] "After the damage to the reputation of a field I cared about, I was very pleased to publish something showing that the idea wasn't crazy," Heeger said. In 2007, Plastic Logic, a company co-founded by Sirringhaus and Sir Richard Friend at Cambridge University, announced plans to build what the company said was the first factory dedicated to making plastic computer chips.

There were scientific surprises for organic electronics too. In 2004, researchers at the University of Manchester in the United Kingdom stumbled across sheets of graphene—planes of carbon atoms in honeycomb structures—that were billions of atoms across and that turned out to be ideal surfaces for on which to run quantum experiments. Scientists working with graphene were able to apply the field effect and measure the quantum Hall effect even at room temperature, a temperature far higher than Schön, constrained by the expectations of known science, had ever dared to claim. The carriers of charge in graphene turned out to be massless particles that moved as fast through the material as light would, a cosmological surprise more profound than anything that Schön had claimed. New avenues of research opened up, triggering physicists' imagination in some of the same ways Schön's claims had done, but they came about through the work of many research groups, working with a variety of lab techniques and materials, not from one person with a single, special recipe.

As science moved on, so did the people. In 2006, Lucent Technologies was taken over by the French communications giant Alcatel. Bell Labs downsized from a research powerhouse that used to publish more than a thousand scientific publications each year to an applied lab that published a couple of hundred in 2007. The Physical Research Lab was reduced to a handful of experts, and former MTSs scattered far and wide, mostly into the physics and engineering departments of universities.

With the dust barely settled from the Schön case, Federico Capasso moved to Harvard University's School of Engineering and Applied Sciences, where he adapted easily to the life of an academic and earned splashes of publicity for work on the Casimir effect, the bizarre quantum mechanical vacuum effect named for the Dutch industrial physicist who coined the idea of the virtuous circle between science and technology. As a professor in the Materials Science and Engineering Department at the University of Illinois at Urbana-Champaign, John Rogers built up a huge research operation producing, among other things, rubber stamps for imprinting circuits onto semiconductor chips, which earned him a flamboyant comparison, by Ira Flatow on National Public Radio's *Science Friday*, to the 1940s Bell Labs transistor pioneers. Cherry Murray, having published her commentary in *Nature Materials* denying that the possibility of fraud in Schön's work had occurred to managers until May 2002, was awarded the American Physical Society George E. Pake prize for management in 2005, in part "for overseeing Bell Laboratories at an important time in its history," having presided over both the unfolding of physics' biggest fraud and its most extensive investigation. She left Bell Labs too, taking a job as the deputy director for Science and Technology at Lawrence Livermore National Laboratory in Livermore, California.

Schön's collaborators also moved on. In 2006, following the formation of Alcatel-Lucent, Christian Kloc left Bell Labs for Singapore.

He had spent a third of his life in Europe, a third in the United States, and it was time to go west again, this time to the Far East, as a professor at Nanyang Technological University, where few people had heard of Hendrik Schön. Zhenan Bao rose to prominence as a chemist and plastic electronics expert, being named as one of *Technology Review*'s TR100 young innovators in 2003, only one year after the Schön case, and earning a professorship at Stanford the year after, suffering no visible damage from her close association with Schön, but being so sensitive about it that she told me she would not answer questions publicly about her role in the case until more than a decade had passed. As for Bertram Batlogg, he continued studying the physics of new materials, including organic materials, at the Swiss Federal Institute for Technology in Zurich, but with a smaller research group than he might otherwise have had. He became involved in a course about ethics in science with a sociology professor and talked about the need to watch out for each small ethical decision, like how many samples had to be studied before a paper could be sent for publication, and the importance of documenting how samples have been prepared. Such small day-to-day decisions, rather than considering the possibility of deliberate fraud, were the main considerations both in doing science and in avoiding becoming the victim of a scientific fraudster, he insisted.

Three years after the Schön case, I was invited to a workshop in the field of organic crystals. In one of the presentations, a junior researcher reported on measurements that she had made by attaching metal contacts to the surface of organic crystals. In some crystals, she said, she had increased the voltage between her contacts until she measured free-flowing or "trap-free" charge. This was a claim that several research groups had made following the Schön case. But the junior researcher was interrupted by a questioner, Bertram Batlogg, who was sitting near the front with a letter-size notepad and who voiced the opinion that no one had measured trap-free charge transport by the method described. It was a scientific dispute, and by 2005, the implicit disclaimer that Batlogg no longer stood by any of the claims of trap-free transport made by Schön was more of an unintended side-effect than the purpose of his interruption. A the presentation continued, I noticed a moving arm, belonging to one of the researchers who had spent months trying to replicate Schön's fraudulent results, stretch out up ahead of me. From where I was sitting, it looked as if he curled the fingers on one hand into a momentary fist.

This fist held the answer to my original questions about the self-correcting nature of science. Science was put straight in the Schön case, but only through the actions of individual researchers. From would-be replicators in dozens of labs, to skeptics, a couple of scientists were transformed into whistleblowers. The correcting process turned out to be more human, more haphazard, and far less systematic than the rule that Hendrik Schön was going to keep trying to publish fabrications, any way he could.

And which scientists did the most to contribute to the resolution of the case? Paradoxically, it wasn't those who acted as if they were confident that science was self-correcting, but those who acted as if they were worried that it wasn't. Somewhere, they figured, science was going wrong. Something had been miscommunicated or misunderstood. There was missing information. There hadn't been enough cross-checks. There were problems in the experimental method or the data analysis. People who reasoned in this way were more likely to find themselves pulling on, or uncovering, or pointing out, problems in Schön's work that it appeared might have arisen through clerical errors or experimental artifacts, but that eventually turned out to be the thin ends of wedges supporting an elaborately constructed fraud. Even when they failed to expose fraud, researchers acting in this way were most likely to contribute to the momentum of the eventual whistleblowers, and to begin the task of setting the scientific record straight. In contrast, those who did the most to prolong and sustain the fraud were those who acted as if the self-correcting nature of science could be trusted to come to the rescue, who were confident that future research would fill the gaps. This way of thinking provided a rationalization for publishing papers at journals; both at *Science* magazine, where editors expedited at least one of Schön's papers into print without always following their own review policy, and at *Nature,* where, on at least one occasion, technical questions about Schön's method were less of a priority than accessibility. Bell Labs managers reasoned similarly, taking the decision to put the reputation of a renowned research institution behind research that was known to be puzzling, and so crossed the line between open-mindedness to new results, leading others astray.

It often happens in frontier research. Scientists and managers take a risk on bold new claims, hoping that future science will correct mistakes, and they get away with it. The difference, in a case of fraud, is that the data are tuned to pass checks and fulfill demands in such a way that the natural self-correction processes are confounded at every turn. At Bell Labs, questions about Schön's lab technique that might have been asked, had his data satisfied fewer expectations and ambitions, did not come up until many of his papers had already shipped out. At the journals, review processes, traditionally a strength of science, became opportunities for additional fabrication. Deception, and the evasiveness that went with it, led collaborators and managers to trust not only Schön's integrity, but also his method, his expertise, the technique he was using and the fact that he was using it in the off-site, off-hand way that he was. The support of fraudulent research was unwitting, but the decision to trust, not only a single researcher, but also that myth of science as an inevitably self-correcting process, was a judgment that continues to be defended to this day.

That's where the danger is. Even now, Hendrik Schön would be happy to work in science again. Think, for a moment, about the way he might do so. Somewhere in science, there must be a place for a logical, intelligent and literal-minded student who fits the profile of a solid graduate, but who isn't quite brilliant, who doesn't quite "get it." Schön's gift for agreeing with others while disguising his contrivance to do so, to be plausible and amiable—these qualities won't go away. He will be happy to change fields, to compromise, and to be flexible enough that when science self-corrects, he can self-correct too. A couple of years after this book is printed, his ban by the German Science Foundation will expire. The memories of the impact of the fraud in his community will fade. People won't be clenching fists at conferences any more. Then maybe Schön will find an opening, not at Bell Labs, but, perhaps, in a university research center that is reliant on a corporate sponsor and advancing in a brilliant new research direction, or a government laboratory that needs political favor for a new research initiative, anywhere where the incentive to publish is exacerbated by scientific ambition and economic pressures. If not Schön, there will be others, and would scientific fraudsters less extreme, less literal, and savvier than Schön be less dangerous than he was, or more?

In order to know where Schön, or someone like him, might next show up, and to guard against the possibility that a groundbreaking discovery being reported right now is fraudulent, it helps to keep something in mind: What do scientists want? What method will enable them to study the most interesting problems? What data will show that their experiment is working? What result will help to protect a great laboratory from an unjustified cut in funding? What will change the reviewer's mind? What will see off a competitor or critic? What will take them, and us, far beyond the frustratingly primitive and slow methods and tools of our time, to new horizons? Hold that thing in mind, think maybe also about what you need for your work to succeed, and there he is. Curly brown hair, half a smile and half a shrug, and he's holding out the answer on a printout from nowhere and asking whether there's any follow-up work that would also be useful.

NOTES AND ADDITIONAL REFERENCES

THE NOTES USE THE FOLLOWING ABBREVIATIONS:

"Schön Investigation Report" refers to M. R. Beasley, S. Datta, H. Kogelnick, H. Kroemer, and D. Monroe (2002) Report of the investigation committee on the possibility of scientific misconduct in the work of Hendrik Schön and co-authors. Lucent Technologies, New Jersey. Available at http://publish.aps.org/reports/ doi:10.1003/aps.reports.lucent accessed August 27, 2008.

"Konstanz Report" refers to D. Lorenz, O. Fabel, K. Froböse, G. Seebaß, L. S. Leiderer, and R. Klein (2003) Bericht über die Prüfung wissenschaftlichen Fehlverhaltens von Herrn Dr. Jan Hendrik Schön. University of Konstanz, Germany. Available at http://www.uni-konstanz.de/struktur/schoen.html accessed August 27, 2008.

"Schön's PhD Thesis" refers to J. H. Schön (1997) *Anwendungen von CuGaSe2 in der Photovoltaik.* Dissertation der Universität Konstanz. Band 329, UFO Atelier für Gestaltung und Verlag, Konstanz.

INTRODUCTION

1. Schön Investigation Report.
2. ISI Web of Science search for *Science* and *Nature* papers in 2001 ranked Schön second with eight papers published in the two journals. Charles Lieber of Harvard University, a senior lab head managing a large research group, was top with nine.
3. P. M. Grant (2002) Scientific credit and credibility. *Nature Materials* 1 (3) 139–141.
4. V. Podzorov, S. E. Sysoev, E. Loginova, V. M. Pudalov and M. E. Gershenson (2003) Single-crystal organic field effect transistors with the hole mobility similar to 8 cm(2)/V s. *Applied Physics Letters* 85 (17) 3504–3506. See also V. Podzorov, E. Menard, A. Borissov, V. Kiryukhin, J. A. Rogers and M. E. Gershenson (2004) Intrinsic charge transport on the surface of organic semiconductors. *Physical Review Letters* 93(8)086602.
5. Charles Babbage (1830) *Reflections on the Decline of Science in England and on Some of its Causes.* B. Fellowes, London.
6. C. J. List (1985) Scientific Fraud–Social Deviance Or the Failure of Virtue. *Science Technology & Human Values* 10 (4) 27–36 put forward the same idea:

"if a fraudulent scientist does not at least mimic the current standards and techniques of the discipline, then he or she has no hope whatsoever of success."

7. C. A. Murray and S. R. Das (2003) The price of scientific freedom. *Nature Materials* 2 (4) 204–205.
8. D. Goodstein (2003) In the matter of J. Hendrik Schön. *Physics World* 15 (November) 17.

CHAPTER 1

1. "This is One for Bell Labs" Lucent Technologies' web site, October 1998, at http://www.bell-labs.com/news/1998/october/20/1.html accessed August 27, 2008.
2. Ibid.
3. L. Torsi, A. Dodabalapur, L. J. Rothberg, A. W. P. Fung, and H. E. Katz (1996) Intrinsic transport properties and performance limits of organic field-effect transistors. *Science* 272 (5267) 1462–1464.
4. Y. Y. Lin, D. J. Gundlach, and T. N. Jackson (1996) High mobility pentacene organic thin film transistors. *54th Device Research Conference Digest* 80–81.
5. Alexander Graham Bell (1914) "Discovery and invention." *National Geographic* XXV 649–655.
6. H. Small and D. Pendlebury (1997) Citations of the 1120 most-cited physicists 1981—June 1997. http://physics.bu.edu/~redner/projects/citation/isi.html accessed November 21, 2008.
7. B. Batlogg (2000) Organic molecular crystals: a new playground to study many-body physics. Kavli Institute for Theoretical Physics, University of California Santa Barbara. Recording online at http://online.kitp.ucsb.edu/online/hitc_c00/batlogg/ accessed February 6, 2007.
8. G. Horowitz, F. Garnier, A. Yassar, R. Hajlaoui and F. Kouki (1996) Field-effect transistor made with a sexithiophene single crystal. *Advanced Materials* 8 (1) 52–54.
9. C. Kloc, P. G. Simpkins, T. Siegrist and R. A. Laudise (1997) Physical vapor growth of centimeter-sized crystals of alpha-hexathiophene. *Journal of Crystal Growth* 182 (3–4) 416–427.
10. R. B. Laughlin (2005) *A Different Universe: Re-inventing Physics from the Bottom Down.* Basic Books, New York.

CHAPTER 2

1. My translation. An archived copy of the Bucher lab web site entitled "Open Positions" and dated 1998 is available at http://web.archive.org/web/19990930125833/www.uni-konstanz.de/FuF/Physik/Bucher/stellen.htm accessed August 27, 2008.
2. "Physicists create Organic Electronics and Hi-Speed Circuits" Bell Labs Press Release, March 20, 2000. http://www.bell-labs.com/news/2000/march/20/1.html accessed September 20, 2007.
3. In 2003, a University of Konstanz panel published a report, the Konstanz Report, which concluded that five papers published as part of Schön's PhD work were problematic. Although the report failed to give proper citations for implicated papers, a member of the physics faculty unofficially identified five references, several of which are critiqued in this chapter and in Chapter 3, although often not on the same grounds as they were drawn to the attention of the panel.

4. J. H. Schön, F. P. Baumgartner, E. Arushanov, H. Riazi-Nejad, C. Kloc and E. Bucher (1996) Photoluminescence and electrical properties of Sn-doped CuGaSe2 single crystals. *Journal of Applied Physics* 79 (9) 6961–6965. This paper was apparently cited by the Konstanz Report.
5. Figure 2 in J. H. Schön, F. P. Baumgartner, E. Arushanov, H. Riazi-Nejad, C. Kloc and E. Bucher (1996) *Journal of Applied Physics* 79 (9) 6961–6965. When I repeated Schön's analysis for the data points he shows, I got a linear correlation coefficient smaller than the one he quotes (2.4 versus 2.6). The discrepancy brings his result into line with his cited reference of B. Pődör (1987) On the concentration dependence of the thermal ionization energy of impurities in InP. *Semiconductor Science and Technology* 2 (3) 177–178. There is supporting evidence for a problem with the accuracy of Schön's numbers in that he introduces an unexplained error of 0.2 that changes to 0.5 in his figure caption. His paper was apparently cited by the Konstanz Report.
6. Figure 4 in J. H. Schön, F. P. Baumgartner, E. Arushanov, H. Riazi-Nejad, C. Kloc and E. Bucher (1996) *Journal of Applied Physics* 79 (9) 6961–6965. This paper was apparently cited by the Konstanz Report.
7. J. H. Schön and E. Bucher (1999) Comparison of point defects in CuInSe2 and CuGaSe2 single crystals. *Solar Energy Materials and Solar Cells* 57 (3) 229–237. This paper, apparently cited by the Konstanz Report, reports a result consistent with B. Pődör (1987) *Semiconductor Science and Technology* 2 (3) 177–178, but omits stages in the calculation in J. H. Schön, F. P. Baumgartner, E. Arushanov, H. Riazi-Nejad, C. Kloc and E. Bucher (1996) *Journal of Applied Physics* 79 (9) 6961–6965 that appear problematic.
8. J. H. Schön, F. P. Baumgartner, E. Arushanov, H. Riazi-Nejad, C. Kloc and E. Bucher (1996) *Journal of Applied Physics* 79 (9) 6961–6965 contains a paragraph that is identical to one in S. M. Wasim and G. S. Porras (1983) On the Hole Mobility of CuGaSe2. *Physica Status Solidi A-Applied Research* 79 (1) K65-K68. The lifted section describes part of the analysis used to calculate one of the best fit lines in Figure 4.
9. J. H. Schön, F. P. Baumgartner, E. Arushanov, H. Riazi-Nejad, C. Kloc and E. Bucher (1996) *Journal of Applied Physics* 79 (9) 6961–6965. Schön writes that G. Massé, his reference 11, has found electron energy levels at 50±10 and at 110±10 milli-electron-volts. These appear to be references to Table II in G. Massé (1990) Concerning lattice-defects and defect levels in CuInSe2 and the I-III-IV2 compounds. *Journal of Applied Physics* 68 (5) 2206–2210, which does not include the errors Schön quotes. Schön's data lead him to find that the sum of these two levels is 180 milli-electron-volts, so that the (plausible) errors he attributes allow him to claim agreement with Massé. The paper was apparently implicated by the Konstanz Report.
10. Schön's PhD Thesis, p. 36
11. Figure 2 and page 303 of J. H. Schön, O. Schenker, H. Riazi-Nejad, K. Friemelt, C. Kloc and E. Bucher (1997) Characterization of defect levels in doped and undoped CuGaSe2 by means of photoluminescence measurements. *Physica Status Solidi A-Applied Research* 161 (1) 301–313. The paper was apparently implicated by the Konstanz Report.
12. Figure 3 of J. H. Schön, O. Schenker, L. L. Kulyuk, K. Friemelt and E. Bucher (1998). Photoluminescence characterization of polycrystalline CuGaSe2 thin films grown by rapid thermal processing. *Solar Energy Materials and Solar Cells* 51 (3–4) 371–384. Sample G6 in Figure 3.20 of Schön's PhD Thesis and in Figure 2 of J. H. Schön, O. Schenker, H. Riazi-Nejad, K. Friemelt, C. Kloc

and E. Bucher (1997) *Physica Status Solidi A-Applied Research* 161 (1) 301–313. The data change from Figure 3.30 on p. 48 of the thesis and Figure 2 in the paper in *Solar Energy Materials and Solar Cells.* The papers were apparently among those implicated by the Konstanz Report.

13. The statement echoed Sigmund Freud, who after falling out with his former protégé, Carl Jung, commented that Jung's error in giving a colleague the wrong date for an event might have been unconscious, but was not the unconsciousness of a gentleman. E. Jones (1998) *The Life and Work of Sigmund Freud. Volume II,* Basic Books, New York. p. 145.
14. See, for example, Robert B. Laughlin's autobiography in Tore Frängsmyr Ed. (1999) *Les Prix Nobel. The Nobel Prizes 1998.* Nobel Foundation, Stockholm.
15. My translation. "Open positions" on a copy of the Bucher lab web site is at http://web.archive.org/web/19990930125833/www.uni-konstanz.de/FuF/Physik/Bucher/stellen.htm accessed August 27, 2008.

CHAPTER 3

1. Richard S Westfall (1973) Newton and the fudge factor. *Science* 179 (4075) 751–758.
2. H. W. Turnbull Ed. (1960) *The Correspondence of Isaac Newton Volume II, 1676–1687.* Cambridge University Press for the Royal Society of London. pp. 284, 262, and 183.
3. Alan E. Shapiro (2005) "Skating on the edge: Newton's investigation of chromatic dispersion and achromatic prisms and lenses." in Jed Z. Buchwald and Allan Franklin Eds. (2005) *Wrong for the Right Reasons.* Springer, New York.
4. Isaac Newton (1730) *Opticks: Or, A Treatise of the Reflections, Refractions, Inflections & Colours of Light* 4 London, 1730, p. 112.
5. Alan E. Shapiro (2005) in Jed Z. Buchwald and Allan Franklin Eds. (2005) Springer, New York.
6. Zev Bechler (1975) A less agreeable matter: the disagreeable case of Newton and achromatic aberration. *The British Journal for the History of Science* 8 (2) 101–126.
7. S. Shapin and S. Schaffer (1989) *Leviathan and the Air Pump.* Princeton University Press.
8. as well as from the idea of *virtual witnessing* described by the previous reference, according to which an experiment that is reported in a sufficiently honest and detailed way may not need to be replicated to have been effectively witnessed by the report's readers.
9. Charles Babbage (1830) *Reflections on the Decline of Science in England and on Some of its Causes.* B. Fellowes, London.
10. The Editors of *The New Atlantis* (2006) Rethinking peer review. *The New Atlantis* (13) 106–110.
11. The same idea comes up in D. Goodstein (1997) Conduct and misconduct in science. *Physics World* 10 (March) 13–14. Goodstein quotes an unnamed colleague: "scientific truth is the coin of the realm. Faking data means counterfeiting the coin."
12. Schön's PhD Thesis.
13. J. H. Schön and B. Batlogg (1999) Modeling of the temperature dependence of the field-effect mobility in thin film devices of conjugated oligomers. *Applied Physics Letters* 74 (2) 260–262. The data shown in the figures does not

correspond to the data in the paper's references 13 and 15, from which it is supposedly taken. Tom Jackson said his group noticed the shifting data at the time, and thought it sloppy.

14. J. H. Schön, C. Kloc, R. A. Laudise and B. Batlogg (1998) Electrical properties of single crystals of rigid rodlike conjugated molecules (Retracted article. See vol B 66, art. no. 249904, 2002). *Physical Review B* 58 (19) 12952–12957. Figures 2 and 6 have a data series in common. The captioning is consistent so there is no falsification.
15. Compare the top data series in the bottom panel of Figure 2 in J. H. Schön, C. Kloc, R. A. Laudise and B. Batlogg (1998) *Physical Review B* 58 (19) 12952–12957, submitted in February 1998, and the top series in Figure 1 of J. H. Schön, C. Kloc, R. A. Laudise and B. Batlogg (1998) Surface and bulk mobilities of oligothiophene single crystals. *Applied Physics Letters.* 73 (24) 3574–3576, submitted in September 1998. Three points move and a tail of data has been added.
16. BPA Worldwide's Circulation Report gives a figure of 150,026 for *Science* in June 2000, falling to 130,969 in June 2007.
17. This figure was provided in 2008 by the American Physical Society press office.
18. J. H. Schön and E. Bucher (1999) Comparison of point defects in CuInSe2 and CuGaSe2 single crystals. *Solar Energy Materials and Solar Cells* 57 (3) 229–237 reports a negative result, but Figure 1 of the paper contains an apparent duplication from Figure 1 of J. H. Schön, F. P. Baumgartner, E. Arushanov, H. Riazi-Nejad, C. Kloc and E. Bucher (1996) Photoluminescence and electrical properties of Sn-doped CuGaSe2 single crystals. *Journal of Applied Physics* 79 (9) 6961–6965, an earlier negative result. Both papers were apparently flagged by the Konstanz Report.
19. Biographical information on Jan Hendrik Schön submitted as part of James T. Yardley, Ronald Breslow and Horst Stormer "Columbia Center for Electronic Transport in Molecular Nanostructures." Proposal to the US National Science Foundation, February 15, 2001.
20. J. H. Schön, J. Oestreich, O. Schenker, H. Riazi-Nejad, M. Klenk, N. Fabré, E. Arushanov and E. Bucher (1999) N-type conduction in Ge-doped CuGaSe2. *Applied Physics Letters* 75 (19) 2969–2971. See also the citation to this paper in Y. J. Zhao, C. Persson, S. Lany, and A. Zunger (2004) Why can CuInSe2 be readily equilibrium-doped n-type but the wider-gap CuGaSe2 cannot? *Applied Physics Letters* 85 (24) 5860–5862, and S. Doka Yamigno (2007) *Characterization of as-grown and Ge-ion implanted* $CuGaSe_2$ *thin films prepared by the CCSVT technique.* PhD Dissertation. Free University of Berlin. Doka reports on samples of thin films of CGS treated with Ge-implantation by the same researchers in Toulouse as had treated crystals measured by Schön. Although Doka does not report on work with crystals, his thesis concludes that additional treatments would be necessary to change the charge transport type. Former students in Konstanz were also unable to build on Schön's result in their own work with thin films.
21. J. H. Schön, C. Kloc and B. Batlogg (1999) Reversible gas doping of bulk alpha-hexathiophene. *Applied Physics Letters* 75 (11) 1556–1558.
22. W. Warta and N. Karl (1985) Hot holes in naphthalene–high, electric-field-dependent mobilities. *Physical Review B* 32 (2) 1172–1182.
23. Schön Investigation Report E33.
24. Schön Investigation Report E7. Finding of falsification.

CHAPTER 4

1. J. H. Schön, C. Kloc, R. A. Laudise and B. Batlogg (1998) Surface and bulk mobilities of oligothiophene single crystals. *Applied Physics Letters* 73 (24) 3574–3576. Page 3575 reports on leakage.
2. R. B. van Dover, L. F. Schneemeyer, R. M Fleming (1998) Discovery of a useful thin-film dielectric using a composition-spread approach. *Nature* 392 (6672) 162–164.
3. R. F Service (2002) Bell Labs: Winning streak brought awe, and then doubt. *Science* 297 (5578) 34–37.
4. Schön Investigation Report E8–E10 describe this and reach a finding of falsification.
5. Compare A. Dodabalapur, H. E. Katz, L. Torsi and R. C. Haddon (1996) Organic field-effect bipolar transistors. *Applied Physics Letters* 68 (8) 1108–1110 and J. H. Schön, S. Berg, C. Kloc and B. Batlogg (2000) Ambipolar pentacene field-effect transistors and inverters (Retracted article See vol 298, pg 961, 2002). *Science* 287 (5455) 1022–1023. Dodabalapur, et al, write, "complementary circuits in which a single transistor can function as either an n- or p-channel device have been proposed and analyzed by Neudeck *et al.*, for a-Si based FETs (13)" where ref. 13 is G. W. Neudeck, H. F. Bare and K. Y. Chung (1987) Modeling of Ambipolar A-Si-H Thin-Film Transistors. *IEEE Transactions on Electron Devices* 34 (2) 344–350. Schön et al. write, "complementary circuits in which single transistors operate either as n- or p-channel device [sic] have been proposed and analyzed for a-Si:H-based FETs (22, 23)" where reference 22 is the paper by Neudeck *et al.* Dodabalapur et al's footnote 14 notes the expectation of ambipolarity "in ultrapure organic materials."
6. Schön Investigation Report E25.
7. See the references 1 and 2 to the work of Klaus von Klitzing in J. H. Schön, C. Kloc and B. Batlogg (2000) Fractional quantum Hall effect in organic molecular semiconductors. *Science* 288 (5475) 2338–2340. This paper has been retracted.
8. Michael Riordan and Lillian Hoddeson (1997) *Crystal Fire: The Birth of the Information Age (Sloan Technology Series)* W. W. Norton & Company, New York.
9. R. F. Service (2002) Bell Labs: Winning streak brought awe and then doubt. *Science* 297 (5578) 34–37.
10. J. H. Schön, C. Kloc and B. Batlogg (2000) Efficient organic photovoltaic diodes based on doped pentacene (Retracted article. See vol 422 pg 93 2003). *Nature* 403 (6768) 408–410.
11. J. H. Schön, C. Kloc and B. Batlogg (2000) Ambipolar pentacene field-effect transistors and inverters (Retracted article. See vol 298, pg 961, 2002). *Science* 287 (5455) 1022–1023.
12. B. Batlogg (2000) Organic molecular crystals: a new playground to study many-body physics. Kavli Institute for Theoretical Physics, University of California Santa Barbara. Recording online at http://online.kitp.ucsb.edu/online/hitc_c00/batlogg/ accessed February 6, 2007.
13. T. Siegrist, C. Kloc, J. H. Schön, B. Batlogg, R. C. Haddon, S. Berg and G. A. Thomas (2001) Enhanced physical properties in a pentacene polymorph. *Angewandte Chemie-International Edition* 40 (9) 1732–1736. See also D. Holmes, S. Kumaraswamy, A. J. Matzger and K. P. C. Vollhardt (1999) On the nature of nonplanarity in the [N]phenylenes. *Chemistry-A European Journal* 5 (11) 3399–3412.

14. Reference 11 of T. Siegrist, C. Kloc, J. H. Schön, B. Batlogg, R. C. Haddon, S. Berg and G. A. Thomas (2001) *Angewandte Chemie-International Edition* 40 (9) 1732–1736 is S. Berg, A. Markelz, G. A. Thomas, J. H. Schön, C. Kloc, P. B. Littlewood and B. Batlogg *Physical Review B* (in press), which never appeared in print.
15. M. Lee, J. H. Schön, C. Kloc and B. Batlogg (2001) Electron-phonon coupling spectrum in photodoped pentacene crystals. *Physical Review Letters* 86 (5) 862–865.
16. A. F. Hebard, M. J. Rosseinsky, R. C. Haddon, D. W. Murphy, S. H. Glarum, T. T. M. Palstra, A. P. Ramirez, A. R. Kortan (1991) Superconductivity at 18-K in potassium-doped C–60. *Nature* 350 (6319) 600–601.
17. Schön Investigation Report E28.
18. One of the earliest mentions of insulator breakdown is in J. H. Schön, C. Kloc and B. Batlogg (2000) Superconductivity in molecular crystals induced by charge injection (Retracted article. See vol 422 pg 93 2003). *Nature* 406 (6797) 702–704, submitted April 18, 2000.
19. Schön Investigation Report E7. Finding of falsification.
20. M. Pope and C. E. Swenberg (1999) *Electronic Processes in Organic Crystals and Polymers.* Oxford University Press, Oxford and New York.
21. J. H. Schön, C. Kloc, H. Y. Hwang and B. Batlogg (2001) Josephson junctions with tunable weak links. *Science* 292 (5515) 252–254. This paper has been retracted.
22. See, for example, E. A. Silinsh and V. Capek (1994) *Organic Molecular Crystals: Interaction, Locatization and Transport Phenomena.* AIP Press, New York. This is cited in J. H. Schön, S. Berg, C. Kloc and B. Batlogg (1999) High mobilities in organic molecular crystals. *Materials Research Society Symposium Proceedings Fall 1999* 598 BB9.5.
23. Schön Investigation Report E34.
24. Stamp said that his concern had been that the mobility of the charge through organic crystals should be proportional to the square of the electric field, which was not what the unpublished Schön data apparently showed. By the time of J. H. Schön, C. Kloc and B. Batlogg (2001) Hole transport in pentacene single crystals (Retracted article. See vol B 66, art. no. 249903, 2002). *Physical Review B* 6324 (24) 245201, Schön and collaborators reported that the mobility was proportional to the square of the field.
25. Ref. 12 of J. H. Schön, S. Berg, C. Kloc and B. Batlogg (1999) High mobilities in organic molecular crystals. *Materials Research Society Symposium Proceedings Fall 1999* 598 BB9.5 is an unpublished manuscript about a crossover between "band-like" and "hopping" charge transport. A manuscript on this crossover appeared in print as J. H. Schön, C. Kloc and B. Batlogg (2001) Universal crossover from band to hopping conduction in molecular organic semiconductors. *Physical Review Letters* 86 (17) 3843–3846. This paper has been retracted.

CHAPTER 5

1. Adjusted Close from $55.40 on 5 January 2000 to $39.55 on 6 January 2000. Yahoo Historical Stock Quotes.
2. "Physicists create Organic Electronics and Hi-Speed Circuits" Bell Labs Press Release, March 20, 2000. http://www.bell-labs.com/news/2000/march/20/1.html. Accessed September 20, 2007.

3. Bertram J. Batlogg, Christian Kloc and Jan Hendrik Schon (sic) (1999) U.S. Patent Application 09/441,751 granted as U.S. Patent 6,284,562 on September 4, 2001.
4. W. F. Brinkman (2003) Integrity in industrial research. *Physics Today* 56 (3) 56–57.
5. "Brinkman Outlines Priorities, Challenges for APS in 2002" *APS News* 11 (1) 3.
6. J. H. Schön, C. Kloc, R. C. Haddon and B. Batlogg. (2000). A superconducting field-effect switch. Science 288 (5466) 656-658. Submitted March 6, 2000 and published April 28, 2000; B. Batlogg, C. Kloc and J. H. Schon (sic) U.S. Patent Application 09/560,729 filed April 28, 2000.
7. Edwin A. Chandross, Brian K. Crone, Ananth Dodabalapur and Robert W. Filas (1999) U.S. Patent 6,136,702 granted October 24, 2000.
8. Bertram J. Batlogg, Christian Kloc and Jan Hendrik Schon (sic) U.S. Patent Application 09/560,729 filed April 28, 2000 was abandoned after the conflict with Chandross et al (1999) U. S. Patent 6, 136, 702, filed November 29, 1999, was pointed out, and the priority was transferred to a new U.S. Patent Application 10/050,729, filed January 16, 2002, which was abandoned after Schön's disgrace.
9. J. H. Schön, S. Berg, C. Kloc and B. Batlogg (2000) Ambipolar pentacene field-effect transistors and inverters (Retracted article. See vol 298, pg 961, 2002). *Science* 287 (5455) 1022–1023; J. H. Schön C. Kloc and B. Batlogg (2001) Ambipolar organic devices for complementary logic. *Synthetic Metals* 122 (1) 195–197.
10. B. Batlogg (2000) Organic molecular crystals: a new playground to study many-body physics. Kavli Institute for Physics, Santa Barbara, August. Recording online at http://online.kitp.ucsb.edu/online/hitc_c00/batlogg/. Accessed February 6, 2007.
11. J. H. Schön, C. Kloc and B. Batlogg (2000) Perylene: A promising organic field-effect transistor material (Retracted Article. See vol 82, pg 1313, 2003). *Applied Physics Letters* 77 (23) 3776–3778.
12. Schön Investigation Report E9.
13. Schön Investigation Report E18-E20.
14. Figures 2 and 5 in J. H. Schön, C. Kloc, A. Dodabalapur and B. Batlogg (2000) An organic solid state injection laser. *Science* 289 (5479) 599–601. This paper has been retracted. See also Schön Investigation Report E18.
15. J. H. Schön, C. Kloc, A. Dodabalapur and B. Batlogg (2000) *Science* 289 (5479) 599–601.
16. Editors at *Science* quoted the median rather than the mean for their average. Half of all papers were accepted slower than the given figure, half faster.
17. L. B. Coleman, M. J. Cohen, D. J. Sandman, F. G. Yamagish, A. F. Garito and A. J. Heeger (1973) Superconducting Fluctuations and Peierls Instability in An Organic Solid. *Solid State Communications* 12 (11) 1125–1132.
18. D. E. Schafer, F. Wudl, G. A. Thomas, J. P. Ferraris and D. O. Cowan (1974) Apparent Giant Conductivity Peaks in An Anisotropic Medium–TTF-TTNQ. *Solid State Communications* 14 (4) 347–351.
19. G. A. Thomas, D. E. Schafer, F. Wudl, P. M. Horn, D. Rimai, J. W. Cook, D. A. Glocker, M. J. Skove, C. W. Chu, R. P. Groff, J. L. Gillson, R. C. Wheland, L. R. Melby, M. B. Salamon, R. A. Craven, G. Depasquali, A. N. Bloch, D. O. Cowan, V. V. Walatka, R. E. Pyle, R. Gemmer, T. O. Poehler, G. R. Johnson, M. G. Miles, J. D. Wilson, J. P. Ferraris, T. F. Finnegan, R. J. Warmack, V. F. Raaen and D. Jerome (1976) Electrical-Conductivity of Tetrathiafulvalenium-Tetracyanoquinodimethanide (TTF-TTNQ). *Physical Review B* 13 (11) 5105–5110.

20. Capasso was commenting on an article of mine, E. S. Reich (2005) "Crazy About Crystals" *New Scientist,* March 19, 2005, 185 (2491) 38–42, which described the fact that some researchers inspired by Schön's claims went on to do productive work using different techniques. Capasso called this "an interesting hypothesis" but went on to express the concern that it could give young people the idea that there was nothing wrong with cheating.
21. E. S. Reich (2005) "Crazy About Crystals" *New Scientist,* March 19, 2005, 185 (2491) 38–42.
22. "First Electrically Powered Organic Laser May Lead to More Widespread Use of Lasers for Various Applications" Lucent Press Release, June 28, 2000. A version of this release is available at http://findarticles.com/p/articles/mi_m0EIN/is_2000_July_28/ai_63733642 accessed August 27, 2008.
23. Compare Figures 2 and 5 in J. H. Schön, C. Kloc, A. Dodabalapur and B. Batlogg (2000) *Science* 289 (5479) 599–601 with the insets in Figure 7 of J. H. Schön, A. Dodabalapur, C. Kloc and B. Batlogg (2000) Organic single crystals for electronic and optoelectronic devices. *Materials Research Society Symposium Proceedings Fall 2000* 660 JJ2.1. The data also appears in Figure 5 of J. H. Schön (2002) "Organic Semiconductor single-crystal devices" in V. M. Agranovich and G. C. La Rocca Eds. (2002) *Organic Nanostructures: Science and Applications. Proceedings of the 'Enrico Fermi' Summer School, Volume 149, Varenna, 31 July–10 August 2001.* IOS Press.
24. A. Dodabalapur, H. E. Katz, and L. Torsi (1996) Molecular orbital energy level engineering in organic transistors. *Advanced Materials* 8 (10) 853–855. This is a research news article.
25. J. H. Schön, A. Dodabalapur, C. Kloc and Batlogg, B. (2000) A light-emitting field-effect transistor. *Science* 290 (5493) 963-965. This paper has been retracted.
26. Kimberly Patch "Cheaper Lasers On the Way" *Technology Research News,* 13 September 2000. http://www.trnmag.com/Stories/091300/Organic_Lasers.htm accessed October 23, 2007.
27. M. A. Baldo, R. J. Holmes and S. R. Forrest (2002) Prospects for electrically pumped organic lasers. *Physical Review B* 66 (3) 035321.

CHAPTER 6

1. *Nature* is owned by the same company as Palgrave Macmillan, this book's publisher. I have written for *Nature*'s news section.
2. BPA Worldwide Circulation Report quotes *Nature*'s circulation as 61,220 in June 2000, falling to 56,067 in June 2007.
3. These priorities were reversed in 2000. See "Mission : company information : about NPG" at http://www.nature.com/npg_/company_info/mission.html accessed December 3, 2008.
4. As *Science* did in 2006, following revelations of irregular data in papers by South Korean cloning pioneer Woo Suk Hwang.
5. "What is AAAS" http://www.aaas.org/about accessed December 2, 2008.
6. AAAS Form 990 (1999, 2000, 2001, 2002) show that AAAS Science Publications, which handles advertising for *Science,* is for-profit. It is in turn owned by AAAS. AAAS is non-profit, but pays tax on "unrelated business income" from advertising revenues, according to information from the AAAS press office. AAAS press officer Ginger Pinholster became very defensive when asked to explain whether this means that *Science* is partly profit-making. The substance of an explanation forwarded from AAAS Chief Financial Officer, Colleen Struss, was that revenues from AAAS Science Publications are spent

on AAAS programs in accordance with the AAAS mission. Under U.S. tax law, however, unrelated business income is due when an income source is "not substantially related" to the exempt purpose of an organization.

7. D. Kennedy (2006) The mailbag. *Science* 311 (5765) 1213.
8. Schön Investigation Report E58.
9. J. H. Schön, C. Kloc, H. Y. Hwang and B. Batlogg (2001) Josephson junctions with tunable weak links. *Science* 292 (5515) 252–254.
10. Schön Investigation Report E58 says that a referee complained about the magnitude of the current prior to publication but was ignored. I believe that this is incorrect. Schön's original manuscript showed current in arbitrary units, and the paper only went through one round of review. Although, as the text explains, the referee raised several relevant questions, I believe that the criticism concerning the magnitude of the current was first formulated after publication.
11. J. H. Schön, C. Kloc and B. Batlogg (2000) Superconductivity in molecular crystals induced by charge injection (Retracted article. See vol 422 pg 93 2003). *Nature* 406 (6797) 702–704.
12. Schön Investigation Report E36–E38.
13. J. H. Schön, C. Kloc and B. Batlogg (2001) Ballistic hole transport in pentacene with a mean free path exceeding 30 μm (Retracted article. See Vol 90, pg 3419, 2001). *Journal of Applied Physics* 90 (7) 3419–3421.
14. In 1996, a reviewer for the journal *Oncogene* questioned a falsified figure in a manuscript from the laboratory of Francis Collins, director of the National Center for Human Genome Research at the National Institutes of Health. Collins went on to find fraud in five papers by his graduate student, Amitov Hajra. See C. Macilwain (1996) 'Ambition and impatience' blamed for fraud. *Nature* 384 (6604) 6–7.
15. J. H. Schön, A. Dodabalapur, Z. Bao, C. Kloc, O. Schenker and B. Batlogg (2001) Gate-induced superconductivity in a solution-processed organic polymer film (Retracted article. See vol 422 pg 92 2003). *Nature* 410 (6825) 189–192.
16. J. H. Schön, C. Kloc and B. Batlogg (2000) Superconductivity in molecular crystals induced by charge injection. *Nature* 406 (6797) 702–704. This article has been retracted.
17. J. H. Schön, C. Kloc and B. Batlogg (2001) High-temperature superconductivity in lattice-expanded C–60. *Science* 293 (5539) 2432–2434; J. H. Schön, M. Dorget, F. C. Beuran, X. Z. Xu, E. Arushanov, M. Lagues and C. D. Cavellin (2001) Field-induced superconductivity in a spin-ladder cuprate. *Science* 293 (5539) 2430–2432. Both articles have been retracted.
18. Editors at *Science* told me they use the median rather than the mean as a measure of acceptance times, so I have done the same. As they put it, half of papers were accepted quicker than the median, and half slower.
19. "Science Express" http://www.sciencemag.org/feature/express/introduction.dtl #link1 accessed June 1, 2008.
20. This paper was apparently accepted for publication, but never appeared in print, because Schön was asked to withdraw all submissions after public fraud allegations were made, and did so. I was not able to obtain a copy of the submission, but have seen email discussions about it.
21. S. Kim, G. S. Lee, S. H. Lee, H. S. Kim, Y. W. Jeong, J. H. Kim, S. K. Kang, B. C. Lee and W. S. Hwang (2005) Embryotropic effect of insulin-like growth factor (IGF)-I and its receptor on development of porcine preimplantation embryos produced by in vitro fertilization and somatic cell nuclear transfer. *Molecular Reproduction and Development* 72 (1) 88–97.

22. P. A. De Sousa and I. Wilmut (2007) Human parthenogenetic embryo stem cells: appreciating what you have when you have it. *Cell Stem Cell* 1 (3) 243–244; K. Kim, K. Ng, P. J. Rugg-Gunn, J. H. Shieh, O. Kirak, R. Jaenisch, T. Wakayama, M. A. Moore, R. A. Pedersen and G. Q. Daley (2007) Recombination signatures distinguish embryonic stem cells derived by parthenogenesis and somantic cell nuclear transfer. *Cell Stem Cell* 1 (3) 346–352.
23. W. S. Hwang, Y. J. Ryu, J. H. Park, E. S. Park, E. G. Lee, J. M. Koo, H. Y. Jeon, B. C. Lee, S. K. Kang, S. J. Kim, C. Ahn, J. H. Hwang, K. Y. Park, J. B. Cibelli and S. Y. Moon (2004) Evidence of a pluripotent human embryonic stem cell line derived from a cloned blastocyst. *Science* 303 (5664)1669–74; D. Kennedy (2006) Response to "Committee Report. *Science* Supplementary Information to *Science* 314 (3804) 1353 DOI: 10.1125/science1137840 at www.sciencemag.org/cgi/content/full/314/5804/1353/DC1/1 accessed December 30, 2008. p. 7. Kennedy admits that figure 4 of Hwang's paper was added during the review process.
24. J. Brauman, J. Gearhart, D. Melton, L. Miller, L. Partridge and G. Whitesides (2006) Committee Report. *Science* Supplementary Information to *Science* 314 (3804) 1353; DOI: 10.1125/science1137840 at www.sciencemag .org/cgi/content/full/314/5804/1353/DC1/1 accessed December 30, 2008. p. 4.
25. N. Wade "New criticism rages over South Korean cell research." *New York Times,* December 10, 2006, A6; E. S. Reich "Stem cell pioneer's findings in doubt." *New Scientist,* December 14, 2005, 188 (2530) 6–7.
26. N. R. Alpert, R. C. Budd, L. D. Haugh, C. G. Irvin, W. W. Pendlebury (2002) In the matter of Eric T. Poehlman. Investigation Report, University of Vermont, Burlington, Vermont. Available at http://ori.hhs.gov/documents/uvm _report.pdf accessed April 28, 2008, p. 18.
27. See also M. K. Richardson and G. Keuck (2001) A question of intent: when is a 'schematic' illustration a fraud? *Nature* 410 (6825)144.
28. Schön Investigation Report E5-E7 concluded that Figure 2 in the paper J. H. Schön, H. Meng, and Z. Bao (2001) Self-assembled monolayer organic field-effect transistors. *Nature* 413 (6857) 713–716 was a manipulated falsification of two figures from two earlier papers on organic crystals.
29. Schön Investigation Report E5-E7 found that Figure 3 was a manipulated version of Figure 2 in J. H. Schön, H. Meng and Z. Bao (2001) Self-assembled monolayer organic field-effect transistors (Retracted article. See vol 422 pg 92 2003). *Nature* 413 (6857) 713–716.
30. R. F. Service (2001) Molecules get wired. *Science* 294 (5551) 2442–2443.
31. Schön Investigation Report E52.

CHAPTER 7

1. C. Weber, C. Kloc, M. Martin, J. H. Schoen [*sic*] B. Batlogg and J. Orenstein (2001) Infrared conductivity of photocarriers in organic molecular crystals. *Advanced Light Source.* Lawrence Berkeley National Lab Abstract at http:// www-als.lbl.gov/als/compendium/AbstractManager/uploads/01024.pdf accessed August 27, 2008.
2. C. D. Frisbie (2001) Organic solids hit the fast lane. *Chemical and Engineering News* 79 (13). http://pubs.acs.org/cen/125th/pdf/7913frisbie.txt.pdf accessed February 15, 2009.

3. This number also appeared in the manuscript J. H. Schön (May 2002) Sputtering of alumina thin fields for field-effect doping. Unpublished.
4. R. J. Chesterfield, C. R. Newman, T. M. Pappenfus, P. C. Ewbank, M. H. Haukaas, K. R. Mann, L. L. Miller and C. D. Frisbie (2003) High electron mobility and ambipolar transport in organic thin-film transistors based on a pi-stacking quinoidal terthiophene. *Advanced Materials* 15 (15) 1278.
5. K. A. Parendo, K. H. Sarwa, B. Tan, A. Bhattacharya, M. Eblen-Zayas, N. E. Staley and A. M. Goldman (2005) Electrostatic tuning of the superconductor-insulator transistion in two dimensions. *Physical Review Letters* 94 (19) 197004.
6. C. H. Ahn, A. Bhattacharya, M. Di Ventra, J. N. Eckstein, C. D. Frisbie, M. E. Gershenson, A. M. Goldman, I. H. Inoue, J. Mannhart, A. J. Millis, A. F. Morpurgo, D. Natelson and J. M. Triscone (2006) Electrostatic modification of novel materials. *Reviews of Modern Physics* 78 (4) 1185–1212.
7. E. S. Reich "Crazy About Crystals" *New Scientist,* March 19, 2005, 185 (2491) 38–42; R. W. I. de Boer, T. M. Klapwijk, A. F. Morpurgo (2003) Field-effect transistors on tetracene single crystals. *Applied Physics Letters* 83 (21) 4345–4347.
8. R. W. I. de Boer, A. F. Stassen, M. F. Cracium, C. L. Mulder, A. Molinam, S. Rogge and A. F. Morpurgo (2005) Ambipolar Cu- and Fe-phthalocynanine single crystal field-effect transistors. *Applied Physics Letters* 86 (26) 262109.
9. V. Podzorov, V. M. Pudalov and M. E. Gershenson (2003) Field-effect transistors on rubrene single crystals with parylene gate insulator. *Applied Physics Letters* 82 (11) 1739–1741.
10. B. Batlogg (2000) Organic Molecular Crystals: a new playground to study many-body physics. Kavli Institute for Theoretical Physics, University of Santa Barbara, August. Recording online at http://online.kitp.ucsb.edu/online/hitc_c00/batlogg/ accessed February 6, 2007.
11. T. Ishiguro "Superconductivity by Field Effect Doping" *Superconductivity Web 21,* Spring 2002, p. 11. Available at http://www.istec.or.jp/Web21/PDF/Past-pdf/E-pdf/02_Spring.pdf accessed July 2, 2008.
12. B. Batlogg (2000) Organic Molecular Crystals: a new playground to study many-body physics. Kavli Institute for Physics, Santa Barbara, August. Recording online at http://online.kitp.ucsb.edu/online/hitc_c00/batlogg/ accessed February 6, 2007.
13. R. B. Laughlin (2002) Truth, ownership, and the scientific tradition. *Physics Today* 55 (12) 10–11. Laughlin summarizes the Schön case as having involved "the ostensibly false claims of a new technology to suppress avalanche breakdown in field-effect transistors."
14. Schön Investigation Report E27.
15. S. Yunoki and G. Sawatzky (2001) Absence of magic fillings of carrier-doped C–60 in a field-effect transistor. *arXiv* cond-mat/0110602v1 at xxx.lanl.gov/abs/cond-mat/110602v1 accessed December 30, 2008.
16. J. H. Schön (November 2001) Sputtering of alumina thin films for field effect doping. Unpublished.

CHAPTER 8

1. C. A. Murray and S. R. Das (2003) The price of scientific freedom. *Nature Materials* 2 (4) 204–205 suggests that Schön's trips to Konstanz began at the time of the visa delay in 2001. In fact, Schön traveled back to Konstanz to do experiments as far back as 1999. Of thirty papers that were eventually re-

tracted, Schön had submitted eighteen prior to the visa delay, including three that were press-released by Lucent and twelve after.

2. On July 11, 2003, the *Stuttgarter Zeitung* reported that the site was showing simply empty beer bottles.
3. J. H. Schön (2001) New phenomena in high mobility organic semiconductors. *Physica Status Solidi B-Basic Research* 226 (2) 257–270. This review paper contained a mini-biography that described Schön as having graduated with a diploma on sputtering metal/dielectric multi-layers in 1993.
4. J. H. Schön (2002) "Organic Semiconductor single-crystal devices" in V. M. Agranovich and G. C. La Rocca Eds. (2002) *Organic Nanostructures: Science and Applications. Proceedings of the International School of Physics "Enrico Fermi," Varenna on Como Lake, 31 July–10 August 2001.* IOS Press.
5. J. H. Schön (May 2002) Sputtering of alumina thin fields for field-effect doping. Unpublished. This manuscript circulated in several versions.
6. Or Schön would not have asked Kloc about the possibility of staking a claim on just one sample. See Chapter 5.
7. J. H. Schön (2001) "Plastic fantastic: electronic and optoelectronic devices based on organic materials" Laboratory Robotics Interest Group Mid-Atlantic Chapter, November 8, 2001 at http://lab-robotics.org/Mid_Atlantic/meetings/0111.htm accessed January 4, 2009.
8. Jason Krause (2001) "Blinded with Science" *The Industry Standard,* March 16, 2001. Available online at http://findarticles.com/p/articles/mi_m0HWW/is_12_4/ai_72886892 accessed November 14, 2008.
9. Schön Investigation Report E21-E24.
10. R. E. Dinnebier, O. Gunnarsson, H. Brumm, E. Koch, P. W. Stephens, A. Hug and M Jansen (2002) Structure of haloform intercalated C–60 and its influence on superconductive properties. *Science* 296 (5565) 109–113.
11. J. H. Schön, M. Dorget, F. C. Beuran, X. Z. Xu, E. Arushanov, M. Lagues, C. Deville Cavellin (2001) Field-induced superconductivity in a spin-ladder cuprate. *Science* 293 (5539) 2430–2432.
12. Charles Babbage (1830) *Reflections on the Decline of Science in England and on Some of its Causes.* B. Fellowes, London.
13. "Bell Labs Forscher in Aller Welt" (2002) *Lux* 6 p. 34. Lucent Technologies' Bell Labs, Germany.
14. B. Batlogg and C. Varma (2000) The underdoped phase of cuprate superconductors. *Physics World* 13 (February) 33.

CHAPTER 9

1. E. Drexler (1987) *Engines of Creation: The Coming Era of Nanotechnology.* Anchor Press, New York.
2. A. Aviram and M. A. Ratner (1974) Molecular rectifiers. *Chemical Physics Letters* 29 (2) 277–283.
3. M. A. Reed, C. Zhou, C. J. Muller, T. P. Burgin and J. M. Tour (1997) Conductance of a molecular junction. *Science* 278 (5336) 252–254.
4. J. Chen, M. A. Reed, A. M. Rawlett and J. M. Tour (1999) Large on-off ratios and negative differential resistance in a molecular electronic device. *Science,* 286 (5444) 1550–1552.
5. C. P. Collier, E. W. Wong, M. Belohradsky, F. M. Raymo, J. F. Stoddart, P. J. Kuekes, R. S. Williams and J. R. Heath (1999) Electronically configurable molecular-based logic gates. *Science* 285 (5426) 391–394.

6. R. F. Service (2003) Molecular electronics—next-generation technology hits an early midlife crisis. *Science* 302 (5645) 556–558; More on molecular electronics. *Science* 303 (5661) 1136–1137.
7. Cherry A. Murray, Interview on National Academies of Sciences Web Site. Recording Online at http://www.nasonline.org/site/PageServer?pagename =INTERVIEWS_Cherry_Murray accessed June 5, 2008.
8. V. Jamieson The industrial physicist who has it all. *Physics World* 14 (May) 9.
9. James T. Yardley, Ronald Breslow and Horst Stormer "Columbia Center for Electronic Transport in Molecular Nanostructures." Proposal to the U.S. National Science Foundation, February 15, 2001; "Brookhaven Center for Functional Nanomaterials." Proposal to the U.S. Department of Energy, November 2001.
10. Mark A. Reed and James M. Tour "Computing with Molecules" *Scientific American,* June 2000, 282 (6) 86–93.
11. C. Joachim, J. K. Gimzewski, and A. Aviram (2000) Electronics using hybrid-molecular and mono-molecular devices. *Nature* 408 (6812) 541–548.
12. B. Batlogg (2000) Organic molecular crystals—a new playground to study many-body physics. Kavli Institute for Theoretical Physics, University of Santa Barbara, August 2000. Recording online at http://online.kitp.ucsb.edu/online/hitc_c00/batlogg/ accessed February 6, 2007.
13. L. Bruno "Inside the invention factory." *Red Herring* July 2002, (115) 54–59.
14. J. H. Schön and Z. Bao (2002) Nanoscale organic transistors based on self-assembled monolayers (Retracted Article. See vol 82, pg 1313, 2003). *Applied Physics Letters* 80 (5) 847–849.
15. Greg Kochanski (2005) Performance reviews from the inside. *Oxford Magazine,* Fourth Week Trinity Term 2005, 239 5. Also at kochanski.org/gpk/misc/2005/perf_review.html accessed December 30, 2008. As well as giving ballpark dollar amounts, this article, by a former Bell Labs MTS, describes the deterioration of management-staff relations.
16. This count refers to permanent staff listed on internal organizational charts from 1997 and 2001. The number of non-permanent staff such as visiting scientists and postdocs actually rose slightly over the same period, but the rise did not make up for the fall. A. Michael Noll (2003) Telecommunication basic research: an uncertain future for the Bell Legacy. *Prometheus* 21 (2) 178–193 records a drop in headcount from 1200 in 1997 to 550 in 2001.
17. Bell Labs Product Announcement October 17, 2001. Archived by the Way Back Machine at http://web.archive.org/web/20011109042545/http://www.lucent.com/pressroom/webcast/ dated November 9, 2001 and accessed April 25, 2007.
18. "Bell Labs scientists usher in new era of molecular-scale electronics" Bell Labs Press Release October 17, 2001. Archived by the press web service Eurekalert at http://www.eurekalert.org/pub_releases/2001-10/ltl-bls101501.php accessed October 6, 2006.
19. James T. Yardley, Ronald Breslow and Horst Stormer "Columbia Center for Electronic Transport in Molecular Nanostructures." Proposal to the US National Science Foundation, February 15, 2001. Although Störmer never co-authored with Schön, this proposal states that Störmer and three other named professors "will collaborate with Dr Schön and other staff at Lucent" to make field-effect transistors. The grant was funded in September 2001.
20. J. H. Schön and Z. Bao (2002) *Applied Physics Letters* 80 (5) 847–849.
21. Schön Investigation Report D7, E57.
22. This may be true. Schön Investigation Report E47 stops short of a misconduct finding for a case in which Schön presented, but did not publish, fabricated data.

23. Form 8K Current Report Lucent Technologies, Securities and Exchange Commission Filing November 7, 2001. Available at http://www.sec.gov/Archives/edgar/data/1006240/000095012301507945/y54618e8-k.htm accessed November 13, 2008.
24. C. A. Murray and S. R. Das (2003) The price of scientific freedom. *Nature Materials* 2 (4) 204–205 claimed that Schön was the lab's fourth most prolific writer in 2001. My ISI Web of Science search ranked Schön above all other Bell Labs researchers, with 39 research articles in 2001.
25. Kitta Macpherson "A Trailblazing Star Arises from Bell Labs." *Star Ledger,* February 24, 2002, p. 17.

CHAPTER 10

1. S. Shapin (1994) *A Social History of Truth: Civility and Science in Seventeenth-Century England (Science and Its Conceptual Foundations series).* University of Chicago Press.
2. Harry Collins (2004) *Gravity's Shadow: The Search for Gravitational Waves.* University of Chicago Press, p. 401.
3. J. Evans (1996) Fraud and illusion in the anti-Newtonian rear guard: the Coultaud-Mercier affair and Bertier's experiments, 1767–1777. *Isis,* 87 (1) 74–107.
4. F. Ventura (1990) Grand Master de Rohan's Astronomical Observatory (1783–1789). *Melita Historica. New Series.* 10 (4) 245–255.
5. L. Patkós (2004) The Pasquich affair. *The European Scientist, Symposium on the era and work of Franz Xaver von Zach (1754–1832) Proceedings of the Symposium held in Budapest on September 15–17, 2004. Acta Historica Astronomiae* 24 pp. 182–187.
6. D. Kevles (1998) *The Baltimore Case: A Trial of Politics, Science and Character.* W. W. Norton & Company; H. F. Judson (2004) *The Great Betrayal: Fraud in Science.* Harcourt.
7. D. Weaver, M. H. Reis, C. Albanese, F. Constantini, D. Baltimore and T. Imanishi-Kari (1986) Altered repertoire of endogenous immunoglobulin gene-expression in transgenic mice containing a rearranged mu heavy chain-gene. *Cell* 45 (2) 247–259. Retracted by D. Weaver, C Albanese, F Constantini and D Baltimore (1991) *Cell* 65 (4) 536.
8. The Department of Health and Human Services (1996) Thereza Imanishi Kari PhD. DAB No. 1582. Available at http://www.hhs.gov/dab/decisions/dab1582.html accessed July 31, 2008.
9. See counts 1 and 2 under Federal Register January 23, 2009, 74 (14) 4201–4202.
10. E.S. Reich "Scientific misconduct report under wraps" *New Scientist,* November 25, 2007, 196 (2631) 16.
11. E.S. Reich "MIT professor sacked for fabricating data" New Scientist, October 28, 2005, at http://www.newscientist.com/article.ns?id=dn8230&feedId=online-news_rss091 accessed October 28, 2005.
12. Office of Research Integrity, U.S. Dept of Health and Human Services (1993) Position Paper #1 on "Protection for whistleblowers" http://ori.hhs.gov/misconduct/Whistleblower_Privilege.shtml accessed July 24, 2008.
13. I first heard the formulation that institutional officials are susceptible to "intuitive mistakes" from C. K. Gunsalus. See E. S. Reich (2007) *Nature* 447 (7142) 238–9.
14. Schön Investigation Report D10. In addition to J. H. Schön, H. Meng, and Z. Bao (2001) Self-assembled monolayer organic field-effect transistors. *Nature*

413 (6857) 713–716, I noticed the sentence in a pre-publication draft of J. H. Schön, H. Meng, Z. N. Bao (2001) Field-effect modulation of the conductance of single molecules. *Science* 294 (5549) 2138–2140. It did not appear in the published version of the latter paper. Both papers were retracted.

15. In Harry Collins (2004) *Gravity's Shadow: The Search for Gravitational Waves.* University of Chicago Press, p. 164. Collins describes how critics of controversial claims to have detected gravitational waves circulated transcripts of Langmuir's lecture in the community of gravitational wave researchers. I was reminded of this when a scientist who intended to critique wishful believers in Schön's SAMFET directed me to an online transcript of the lecture, which Collins cites as I. Langmuir (1989) *Physics Today* 42 (10) 36–48.
16. J. H. Schön, H. Meng, Z. N. Bao (2001) Field-effect modulation of the conductance of single molecules. *Science Express* doi:10.1126/science.1066171. Published as *Science* 294 (5549) 2138–2140, and later retracted.
17. Cherry A. Murray and Saswato R. Das (2003) The price of scientific freedom. *Nature Materials* 2 (4) 204–205.
18. Form 10-K Lucent Technologies, Securities and Exchange Commission Filing for the year ended September 30, 2001; 2000 Annual Report, Lucent Technologies, p. 3. The second brochure was not filed at the SEC but is archived at http://www.porticus.org/bell/att/historical_financial.htm accessed December 3, 2008.
19. This prize was known as the Otto-Klung Prize until 2001, and then renamed the Otto-Klung-Weberbank Prize. The prize amount was DM 50,000 in 2001.
20. "Nobelpreisverdächtig" *FU-Nachrichten* 1/2002. Archived at http://web.archive.org/web/20080107100641rn_1/web.fu-berlin.de/fun/2002/01–02/leute/leute4.html accessed original site May 14, 2008.
21. Cherry A. Murray and Saswato R. Das (2003) *Nature Materials* 2 (4) 204–205.
22. Cherry A. Murray and Saswato R. Das (2003) *Nature Materials* 2 (4) 204–205.
23. Zhenan Bao, John A. Rogers and Jan Hendrik Schon (*sic*) (2001) U.S. Patent Application 09/860,107, abandoned after Schön's disgrace.
24. Zhenan Bao, Robert William Filas, Peter Kian-Hoon Ho and Jan Hendrik Schon (*sic*) (2001) U.S. Patent Application 09/951,055 abandoned after Schön's disgrace.
25. J. H. Schön and Z Bao (2002) Nanoscale organic transistors based on self-assembled monolayers. *Applied Physics Letters* 80 (5) 847–849.
26. Horst L. Störmer (2000) The Bell Labs—conditions for basic research at privately financed institutions. *Innovative Structures in Basic Research Conference of the Max Planck Society, Ringberg Castle, Germany,* p. 107.
27. Schön Investigation Report p. 11.
28. J. H. Schön, M. Dorget, F. C. Beuran, X. Z. Zu, E. Arushanov, C. D. Cavellin and M. Laguës (2001) Superconductivity in CaCuO2 as a result of field-effect doping (Retracted article. See vol 422 pg 92 2003). *Nature* 404 (6862) 434–436. Figure 5.
29. Schön Investigation Report E32; J. H. Schön, M. Dorget, F. C. Beuran, X. Z. Zu, E. Arushanov, C. D. Cavellin and M. Laguë (2001) *Nature* 404 (6862) 434–436. The finding of fabrication relates to Figure 3.

CHAPTER 11

1. Some scientists speculate that Hendrik Schön misled himself this way. Chapter 2 argues that Schön did not mislead himself so much as encounter confu-

sion about how to analyze data, and Chapters 3 and 4 report that Schön was faking results prior to his first claim of superconductivity. Vladimir Butko clarified his own measurement in a presentation at the American Physical Society March meeting in 2003.

2. C. A. Murray and S. R. Das (2003) The price of scientific freedom *Nature Materials* 2 (4) 204 says that managers asked Schön to send through an erratum.
3. Schön Investigation Report E55; J. H. Schön, S. Berg, C. Kloc and B. Batlogg (2000) Ambipolar pentacene field-effect transistors and inverters. *Science* 287 (5455) 1022–1023. Cherry A. Murray and Saswato R. Das (2003) *Nature Materials* 2 (4) 204 report that this issue was raised by a postdoc. Although the postdoc is not named, John Rogers and his postdoc Lynn Loo were working on the inverter issue independently of Mark Lee. Murray and Das concede that a postdoc had noticed an additional duplication in early May. Although, again, the duplication is not identified, the inverter duplication between Schön's SAMFET papers and his earlier paper on ambipolar pentacene was the second to be noticed by outside scientists, and directly relevant to the inverter issue that John Rogers had asked Mark Lee was asked to examine; it had either been identified internally at Bell Labs, or was about to be.
4. Charles Seife (2008) *Sun in a Bottle: The Strange History of Fusion and the Science of Wishful Thinking.* Viking. This reports on Seife, a reporter at *Science,* receiving a news tip from an editor with information about scientific manuscript.
5. R. F. Service (2002) Pioneering physics studies under suspicion. *ScienceNow* (520) 1 http://sciencenow.sciencemag.org/cgi/content/full/2002/520/1 accessed July 8, 2008.
6. K. Chang "Similar Graphs Raised Suspicions on Bell Labs Research" *New York Times,* May 23, 2002, p. 29.
7. M. Jacoby (2002) Bell Labs papers questioned. *Chemical and Engineering News* 80 (21) 17.
8. Compare Figures 2 and 5 in J. H. Schön, C. Kloc, A. Dodabalapur and B. Batlogg (2000) An organic solid state injection laser. *Science* 289 (5479) 599–601 and Figure 7 in J. H. Schön, A. Dodabalapur, C. Kloc and B. Batlogg (2000) Organic single crystals for electronic and optoelectronic devices. *Materials Research Society Symposium Proceedings Fall 2000 Boston* 660 JJ2.1. The *Science* paper has been retracted.
9. E. G. Emberly and G. Kirczenow (2002) Molecular Spintronics: Spin-Dependent Electron Transport in Molecular Wires. *arXiv* cond-mat/0201344vl at xxx.lanl.gov/abs/cond-mat/0201344v1 accessed January 5, 2009.
10. J. H. Schön, E. G. Emberly and G. Kirczenow (2002) A single molecular spin valve. *Science Express* doi:10.1126/science.1070563 at www.sciencemag.org/cgi/content/abstract/1070563 accessed December 30, 2008. This paper was retracted and never appeared in the print version of the journal.
11. Philip Stamp recalled that the view that a Wigner crystal should form was put to Batlogg by Boris Spivak in December 2000.
12. R. F. Service (2002) Bell Labs: Winning Streak Brought Awe, and Then Doubt. *Science* 297 (5578) 34–37.
13. "2002 MRS Spring Meeting Highlights" at http://www.mrs.org/s_mrs/doc.asp?CID=2162&DID=89490 and http://www.mrs.org/s_mrs/doc.asp?CID=2162&DID=89491 accessed January 6, 2009.
14. "Universität Konstanz entzieht Jan Hendrik Schön den Doktortitel" Press Release 85, July 11, 2004, University of Konstanz at www.uni-konstanz.de/

struktur/service/presse/mittshow.php?nr=85&jj=2004 accessed January 6, 2009.

15. "DFG imposes sanctions against Jan Hendrik Schön" Press Release of the Deutsche Forschungsgemeinschaft, October 2004 at http://www.dfg.de/aktuelles_presse/reden_stellungnahmen/2004/download/ha_jhschoen_1004_en.pdf accessed January 6, 2009.

EPILOGUE

1. E. J. Meijer, D. M. De Leeuw, S. Setayesh, E. Van Veenendaal, B. H. Huisman, P. W. M. Blom, J.C. Hummelen, U. Scherf and T. M. Klapwijk (2003) Solution-processed ambipolar organic field-effect transistors and inverters. *Nature Materials* 2 (10) 678–682.
2. J. Zaumseil, R. H. Friend and H. Sirringhaus (2006) Spatial control of the recombination zone in an ambipolar light-emitting organic transistor. *Nature Materials* 5 (1) 69–74; J. S. Swensen, C. Soci and A. J. Heeger (2005) Light emission from an ambipolar semiconducting polymer field-effect transistor. *Applied Physics Letters* 87 (25) 253511.

ADDITIONAL REFERENCES

William J. Broad and Nicholas Wade (1985) *Betrayers of the Truth.* Oxford University Press, Oxford.

Narain Gehani (2003) *Life in the Crown Jewel.* Silicon Press, Summit, New Jersey.

Jeff Hecht (1999) *City of Light: The Story of Fiber Optics (Sloan Technology Series).* Oxford University Press, USA.

Joseph R. Hixson (1976) *The Patchwork Mouse.* Anchor Press, New York.

Peter Medawar "Is a Scientific Paper a Fraud?" (1963) in *The Strange Case of the Spotted Mouse: and other Classic Essays on Science.* Oxford University Press, Oxford, 1996.

Robert K. Merton (1979) *The Sociology of Science: Theoretical and Empirical Investigations.* University of Chicago Press, Chicago.

Michel de Montaigne "On Liars" in Michel de Montaigne and Jon M. Cohen Eds. (1993) *Essays.* Penguin, London. p 28.

Fritz K. Ringer (1990) *The Decline of the German Mandarins: The German Academic Community 1890–1933.* Wesleyan University Press, Middletown, Connecticut.

Edgar Silinsh (1980) *Organic Molecular Crystals: Their Electronic States.* Springer-Verlag, Berlin-Heidelberg and New York.

Johann Wolfgang von Goethe and Stuart Atkins (1994) *Faust I & II Goethe: The Collected Works, Volume 2.* Princeton University Press, Princeton.

INDEX

A

Abusch-Magder, David, 164, 179
Advanced Light Source (ALS), 130
Agere Systems, 86, 89, 92, 172, 174–176, 178–179, 190–192
Alam, Ashraf, 192–193
Alberts, Vivian, 152
Alcatel, 236, 237
aluminum oxide (as gate insulator), 67–68, 71, 77, 79, 94, 109, 114–116, 127, 129, 133, 135–137, 139–143, 145, 146–150, 153–154, 156, 160, 206, 209–210, 225–226, 231
ambipolarity, 67–68, 93, 101, 133, 136, 141, 235
American Association for the Advancement of Science, 107
American Physical Society
- Batlogg and, 66, 72–73, 92
- Frisbie and, 133
- George E. Pake prize for management, 236
- Oliver E. Buckley Condensed Matter Prize, 203
- promotion of research claims, 189
- Schön and, 58, 81, 133, 209–210
- von Klitzing and, 92

Anderson, Philip, 161
Antoniadis, Dimitri, 193
Applied Physics Letters, 58, 60, 177, 181
Arushanov, Ernest, 31
Avaya, 86
Aviram, Arieh, 164, 165

B

Babbage, Charles, 5, 50–51, 157
Baldo, Marc, 102, 149
Baltimore case, 187–189
Bao, Zhenan
- Capasso and, 201–202
- McEuen and, 223
- Monroe and, 191, 194
- plastic fantastic and, 167–170, 225
- post-scandal, 237
- SAMFET and, 124–125, 175–178, 181, 197–198, 232
- Schön and, 8, 124–125, 154

Bardeen, John, 70, 116
Batlogg, Bertram
- 2000 APS presentation, 143
- American Physical Society and, 209–212
- Bell Labs and, 11–21, 26
- Bhattacharya and, 136
- Braunschweig Award and, 157
- Brinkman and, 153
- Capasso and, 223–224
- "CLOSED" email and, 199
- Goldman and, 134
- Greene and, 205
- high temperature superconductivity and, 17–21, 25
- investigation into experiments, 223–224, 231–233
- Klapwijk and, 139–140
- Kloc and, 33, 43–44
- laser and, 95–96, 99
- Laudise and, 62
- Laughlin and, 147–150
- McEuen and, 220
- move away from Bell Labs, 89–92, 151
- nanotechnology and, 166
- *Nature* magazine and, 63
- Orenstein and, 129–131

Palstra and, 118
post-scandal, 237
resignation from Bell Labs, 89–92, 151
SAMFET and, 198–199
Schön and, 8, 55–59, 111, 113, 124, 127, 157–161
Ziemelis and, 109
Beasley, Malcolm, 224
Bechler, Zev, 47
Bednorz, Georg, 18
Bell Labs
culture of criticism, 56–57, 102, 134, 147, 175, 203, 224–225
culture of hiring, 59–60, 92, 167
culture of management, 2–3, 7–8, 12–17
decline, 74, 91, 170–172, 236
Department of Materials Physics, 13–14, 18, 32, 72, 89–90
Department of Nanotechnology Research, 163–181, 214
Department of Semiconductor Physics Research, 93
Department of Theoretical Physics Research, 16, 78, 116, 175
patents, 12, 85–89, 201, 227
Physical Research Lab, 16, 17, 65, 86–88, 90, 92, 102, 168, 172, 202, 223, 231, 236
post-scandal, 236–239
press releases, 3, 52, 63, 87, 98, 101, 103, 168, 173–175, 180, 190, 192
"stretch goals," 65, 69, 70
Berg, Steffen, 63, 67, 100
Bergemann, Christoph, 158
Bertier, Joseph-Etienne, 185
Bhattacharya, Anand, 133–137
Blumberg, Girsh, 134, 205–206
Bose-Einstein condensation, 108
Boston Marathon, 193
Boyle, Robert, 48–51
Bradford, Monica, 107, 117
Brattain, Walter, 70
Brauman, John, 121
Brinkman, Bill
Batlogg and, 153
Bell Labs and, 65
Crystal Fire and, 70
on journals, 59
laser and, 92–93
McEuen and, 223
retirement from Bell Labs, 167, 172
Schön and, 87–88
Bucher, Ernst
Capasso and, 178
Klapwijk and, 139
Kloc and, 22–23
lab members, 232, 234
samples and, 100, 102
Schön and, 7, 28–34, 40, 42–44, 57, 59–60, 66, 71, 94, 151–152, 158–159, 176, 181

C

Capasso, Federico
Bell Labs and, 93, 95–98
Bucher and, 178–179
Jackson and, 226–227
McEuen and, 223, 232–233
Monroe and, 194–197, 199–201
post-scandal, 236
Rogers and, 194–197, 199–201
SAMFET and, 173–176, 203–207
Schön and, 8, 97–98, 168, 170–171, 229
Slusher and, 95–96
carbon–60, 76, 88, 111, 150, 156–157, 161, 166
Casimir, Hendrik, 16, 236
Central Research Institute of Electric Power Industry (CRIEPI), 159, 209
CGS (copper gallium selenide), 29–31, 34, 152
Chesterfield, Reid, 133–138
City of Paris School of Industrial Physics and Chemistry, 145, 155
"CLOSED" email, 199
Cooper, Leon, 116
copper oxide, 18, 20, 25, 145, 155–156, 206
Corning, 86
Correspondence astronomique, 186
Cotter, Rosalind, 197
Coultaud, Jean, 184–186
Crystal Fire (Riordan and Hoddeson), 70
crystals
Batlogg and, 129–132, 166

Bell Labs and, 20–21, 24–26, 43, 89–90, 92–95, 129–132
Bucher and, 30
carbon–60, 111
CGS, 30
conductivity and, 113–114, 116
data collection on, 65–72
examination of data from, 79–82
growth, 21–24
investigation into experiments with, 146, 148–149, 153–156, 158, 209–211, 219–220
Klapwijk and, 139–142
Kloc and, 23–24, 25–26, 43, 56, 61–63, 74–75
Laudise and, 15, 20–21
Lucent Technologies and, 87–88
organic, 15–17, 20–23, 25, 43, 55, 57, 60–63, 65–69, 72, 74, 78–79, 82, 87–90, 92–94, 100, 111, 113–114, 125, 129, 131–134, 136–137, 139, 142, 146, 149, 153–154, 156, 158, 166, 169, 176, 209–210, 219, 220, 230, 237
quantum Hall effect and, 176
Schön and, 55–56, 58–60, 61–63, 76–79, 99–101, 170
University of Minnesota and, 132–137
Wigner, 230
cuprates. *see* copper oxide

D

Dahlberg, John, 122–123
D'Angos, Jean Auguste, 50, 186
Das, Saswato, 7, 233–234
David Adler Lectureship Award for Material Physics, 72
de Boer, Ruth, 139–141
de Sallo, Denis, 49
Deluc, Jean-Andre, 185
Deville Cavellin, Catherine, 156, 161–162, 206, 233
Dingell, John, 187
Dodabalapur, Ananth, 15, 26, 68, 93–96, 101–102
Dollond, John, 47–48
Drexler, Eric, 163
Dynes, Bob, 142–143

E

Eblen, Melissa, 137
Emberly, Eldon, 229
Encke, Johann Franz, 186
Engines of Creation (Drexler), 163
Euler, Leonhard, 47–48

F

fabrication of data, 2–3, 5–6, 54, 63, 81–82, 90, 94, 113, 119, 121, 125–127, 150, 181, 194, 199, 205, 214, 228, 229, 233, 238–239
falsification of data, 54, 67–68, 77, 91, 148
Federal Policy on Misconduct, 189
Flatow, Ira, 180–181, 236
Forrest, Stephen, 99, 102, 148–149, 221, 230
fractional quantum Hall effect, 12, 24, 69–70, 72, 80, 88, 92, 101, 131, 133, 141–142, 144, 176, 210, 235
Friend, Richard, 235
Frisbie, Dan, 132–136, 138
Furakawa Electric, 86

G

Gammel, Peter, 89
Garfunkel, Art, 210
gate insulators. *see* aluminum oxide; silicon oxide
George E. Pake prize for management, 236
Gerdil, Cardinal Hyacinthe-Sigismond, 186
Gershenson, Mike, 142
Gimzewski, Jim, 165
Goldhaber-Gordon, David, 127, 143–144
Goldman, Allen, 133–134, 136–138
Goldmann, Claudia, 158, 209
Grazul, John, 226
Greene, Laura, 149, 205

H

Haddon, Robert, 76–77, 88
Hall, Edwin, 12
Hall effect. *see* fractional quantum Hall effect
Hamann, Don, 77, 87, 170
Hanson, Brooks, 108–110, 111, 119

Heath, James, 164
Hebard, Art, 143
Heeger, Alan, 97, 235
Hergenrother, Jack, 174, 191
high-temperature superconductivity (high-Tc), 17–21, 160
Hillhouse, Hugh, 138–140
Ho, Peter, 167, 169–170, 175, 178
Holstein, Theodore, 80–81
Hooke, Robert, 46, 47
Hsu, Julia, 173, 176–178, 214–217
Hwang, Harold, 78
Hwang, Woo Suk, 2, 6, 78, 119–121, 123–124

I

Industrial Revolution, 51
Internet boom (bust), 8, 12, 85
investigations
  external (into Schön), 1, 3, 5–6, 8, 121–123, 200–201, 223–227, 232–234
  in general, 31, 67, 76–77, 81–82, 112, 187, 189–190
  internal (into Schön), 2, 198–199, 200, 204, 213–214
Isaacs, Eric, 89–90, 91, 175
Ishiguro, Takehiro, 147

J

Jackson, Tom, 179–180, 226
Jaffe, Jeff, 165, 172
Joachim, Christian, 165–166
Journal des beaux-arts et sciences, 50, 184–185
Journal des sçavans, 49, 50
Journal of Applied Physics, 56, 63–64, 113
Journal of Cell Biology, 53–54
journals (in general), 1–3, 49, 52–53, 59, 105–128. *see also Nature; Science*

K

Karl, Norbert, 62–63, 92, 132, 140
Kekulé, August, 79
Kelly, Stephen, 122
Kennedy, Donald, 107, 110
Kirczenow, George, 229
Klapwijk, Teun, 71, 111–114, 139–141
Kleiman, Rafi, 90, 168, 174–175, 177, 202–204
Klingenstierna, Samuel, 47–48
Kloc, Christian
  Batlogg and, 71–74, 130–131, 148, 150
  Bell Labs and, 21–23, 25–26, 43
  Bucher and, 32–33
  Chesterfield and, 134
  laser and, 96
  McEuen and, 223
  Podzorov and, 142
  post-scandal, 237
  SAMFET and, 175, 176
  samples and, 100
  Schön and, 8, 43, 56, 61–62, 69, 76, 78, 80–81, 89, 127, 151, 154–158, 166, 210–211, 232–233
Kmeth, Daniel, 186–187
Kogelnik, Herwig, 224
Kouwenhoven, Leo, 71, 212–213, 218, 221–222
Kroemer, Herb, 224
Kulyuk, Leonid, 31, 38

L

Laguës, Michel, 145
Langmuir, Irvin, 194
laser (claim by Schön)
  beginnings of, 1, 92–94
  initial data, 92, 94–96
  press release, 98
  reaction to, 98–103, 148, 174
  Slusher and, 95–96
Laudise, Bob, 15, 17, 21–23, 26, 33, 62, 72
Laughlin, Robert, 12, 13, 24, 146–149, 160, 204
Le Sage, George-Louis, 184–186
Lee, Mark, 74, 215–219, 231
Levi, Barbara Goss, 198, 211
Leviathan and the Air Pump (Shapin and Shapiro), 48
Lieber, Charles, 218, 222
Littlewood, Peter, 16, 78, 80–83, 116
Loo, Lynn, 214–216
Los Alamos National Lab, 142, 153–154, 210–212
Lucent Technologies
  Agere Systems, 176

Batlogg and, 157
Bell Labs and, 8, 12–13
loss of revenue, 165, 170–173
patents and, 85–89, 165
plastic electronics and, 16–17, 155
post-scandal, 236–237
SAMFET and, 190–191, 201
Schön and, 59, 65–66, 166, 179–180, 196, 215, 225, 234
Lüscher, Silvia, 144
Lux-Steiner, Martha, 33

M

MacDiarmid, Alan, 97
Marcus, Charles, 218, 222
Markelz, Andrea, 74
Materials Research Society, 76, 99, 132, 180, 230, 234
Max Planck Institute (MPI), 157
McCain, John, 215
McEuen, Paul, 2, 4, 213–214, 218, 219–225, 227, 231–232
McGinn, Richard, 85, 172
McMillan Prize. *see* William L. McMillan Award
Members of Technical Staff (MTS), 13, 15, 23, 26, 43, 65–66, 69–70, 72–74, 78, 85, 89–93, 95, 97, 99–100, 111, 134, 160, 164–181, 191, 194–196, 202, 204–206, 211, 215–218, 226, 229, 232, 236
molecular electronics, 124–126, 163–165, 167–169, 173–174, 179–180, 215, 222, 225
Molecular Reproduction and Development, 225
Monroe, Don, 72, 174–175, 190–204, 220, 223–224, 227
Moore's Law, 164
Morpurgo, Alberto, 113–114, 139–140, 142
Müller, Alexander, 18
Muller, David, 72, 99, 176, 202–203, 206, 225–226
Murray, Cherry
Bell Labs and, 206, 223, 227
first look at data, 168–169
management, 92–93, 165
post-scandal, 236
Schön and, 7, 196, 199–201, 233
"stretch goals" and, 65

N

nanotechnology, 1, 163–181, 212–214. *see also* molecular electronics
Natelson, Doug, 153, 159, 161, 203
National Geographic, 52
Nature magazine
Brauman and, 121
concerns about publication of Schön's work, 169–170, 178, 215, 238
cover on molecular electronics, 166
demand, 131
emphasis on, 59
founding of, 51–52
IBM letter and, 197–198, 201–203
incentive for scientists to publish in, 59
influence on scientific perception, 189–192
Isaacs and, 90
McEuen and, 223–224
publication of Schön's claims, 1, 63, 71, 113, 115–119, 155, 157, 190–192
review of molecular electronics, 166
review of nanotechnology manuscript, 172–173
SAMFET and, 124–128, 172–173, 217–221, 229
Sohn and, 212
submission process, 105–110
web site, 51–52
Nature Materials, 7, 236
Netravali, Arun, 11, 92, 172
Newton, Isaac, 6, 45–48, 50, 188–186
Nobel Prize winners
Baltimore, 187
Batlogg and, 18
Bell Labs and, 12–13
Bose-Einstein condensation, 108
electrically-conducting plastics, 154
fractional quantum Hall effect, 24
Heeger, 97
Kogelnik, 224
Laughlin, 12, 146, 159, 195
Störmer, 11–13, 69–70, 144, 198
Tsui, 12
von Klitzing, 92

n-type materials, 30, 42, 59–61, 67, 100

O

Ocko, Ben, 232
Oldenburg, Henry, 46, 49
Oliver E. Buckley Condensed Matter Prize, 203
Onnes, Heike Kamerlingh, 18
Orenstein, Joe, 129–132, 140, 160, 211
organic materials. *see* crystals; plastics
Osborne, Ian, 111
O'Shea, William, 155, 172, 180
Otto-Klung-Weberbank Prize, 198

P

Palstra, Thomas, 79, 118, 143
Parendo, Kevin, 137
parthenogenesis, 121
Pasquich, Johann, 186–187
Pasquier, Claude, 145
Pflaum, Jens, 62
Philosophical Transactions, 45, 49
Physical Review B, 74, 102
Physical Review Letters, 58, 59
Physics Today, 65, 87, 211
Physics World, 160, 165
Piltdown man, 5
plagiarism, 39, 54, 177
plastics, as semiconductors, 14–15, 15–15, 17, 26, 97, 108
plastics, compared to crystals, 72, 89
plastics, compared to silicon, 61
Podzorov, Vitaly, 137, 142, 145
Poehlman, Eric, 2, 6, 122–123
polarons, 79–84, 236
Pope, Martin, 77–78
press releases, 3, 52, 63, 87, 98, 101, 103, 168, 173–175, 180, 190, 192
Profokiev, Nikolai, 83
p-type materials, 30, 42, 67
publication, role in science, 45–64

Q

quantum Hall effect, 12, 24, 69–70, 72, 77, 80, 88, 92, 101, 131, 133, 141–142, 144, 176, 210, 235

R

Ralph, Dan, 222
Ramirez, Art, 90–91, 99, 142, 153–154, 210–212, 214, 230
Ratner, Mark, 164
record-keeping, 8, 39–41, 44, 81, 233
Reed, Mark, 164, 165, 180
Reichmanis, Elsa, 167, 176, 232
replication efforts
  Bao and, 181
  Bell Labs and, 97, 195, 206, 209–210, 232–233, 238
  Bucher and, 178
  Chesterfield and, 133
  European scientists and, 45–46
  Frisbie and, 133, 138
  Laughlin and, 147, 204
  Levi and, 198
  Los Alamos National Lab and, 142, 153–154, 210
  Minnesota group and, 132–133, 135
  Orenstein and, 132
  Podzorov and, 137
  Ramirez and, 142, 153–154, 210, 230
  rumors of, 143, 160, 210–211, 212, 214
  Störmer and, 153
Rogers, John A.
  Capasso and, 204–207
  challenges to SAMFET and, 214–216, 218–209
  McEuen and, 223–224
  Monroe and, 195–202
  nanotechnology and, 173
  post-scandal, 236
  SAMFET and, 168–169, 181, 229, 232–233
  Schön and, 8, 166–168, 176, 178, 229, 232–233
Rossner, Mike, 53–54
Royal Society of London, 45–46, 48, 51, 183–184
Ruiz, Rosie, 193–194

S

Sage, Leslie, 115
SAMFET (claim by Schön)
  beginnings of, 124–126
  challenges to findings of, 217–221, 224, 228–232
  initial data, 169–170
  McEuen and, 220–221

Monroe and, 190–191, 196–198
press release of data, 174–181
publication of, 190–191
reaction to, 174–181, 193–198, 196–198, 201–204
Sawatzky, George, 71, 83, 149–150, 153
Schacht, Henry, 172, 196
Schenker, Ortwin, 152, 154–155, 160
Schön, Jan Hendrik
American Physical Society and, 58, 81, 133, 209–210
background, 27
Bao and, 8, 124–125, 154
Batlogg and, 8, 55–59, 111, 113, 124, 127, 158–162
Brinkman and, 87–88
Bucher and, 7, 28–34, 40, 42–44, 57, 59–60, 66, 71, 94, 151–152, 158–159, 176, 181
Capasso and, 8, 97–98, 168, 170–171, 229
concerns about publication of work, 169–170, 178, 215, 238
Kloc and, 8, 43, 56, 61–62, 69, 76, 78, 80–81, 89, 127, 151, 166, 210–211, 232–233
Lucent Technologies and, 59, 65–66, 166, 179–180, 196, 215, 225, 234
Nature magazine and, 63, 71
Rogers and, 8, 166–168, 176, 178, 229, 232–233
Schrieffer, Robert, 116
Science Friday, 180, 236
Science magazine
Bell Labs and, 63, 173, 180
claims of fake data and, 212–213, 215, 217–220
emphasis on, 59, 106–107, 131, 189, 194
founding of, 51–52
Hwang, Woo Suk and, 120–124
laser paper, 96, 98
McEuen and, 224–225
Monroe and, 194, 197, 199
physical science coverage, 107–110
publication of Schön's claims, 1–2, 72, 110–112, 118–119, 126–128, 133, 155–158, 168, 229, 235, 238
submissions, 84, 90
Scientific American, 52, 165
self-assembling monolayer field-effect transistor. *see* SAMFET
Shapin, Steven, 48, 50, 183–184
Shapiro, Alan, 46–50
Shockley, William, 70
silicon oxide, 169
Simpkins, Peter, 23
Sirringhaus, Henning, 235
Skeptical Chymist, The (Boyle), 48
Slusher, Dick, 72, 94–96, 98–99, 101, 170, 181
Sohn, Lydia, 2, 4, 212–214, 217–220, 221–222, 225, 227
Solomon, Paul, 197, 202, 203
Spivak, Boris, 83
Stamp, Philip, 83
Störmer, Horst, 11–13, 70, 75, 144–145, 153, 175, 195, 198–199, 204, 207, 232
superconductivity, high-temperature (high-Tc), 18–20, 160
Swenberg, Charles E., 77–78
Szuromi, Phillip, 117, 225

T

Takeya, Jun, 159, 219
Tan, Sarwa, 137
Tanigaki, Katsumi, 143
Thomas, Gordon, 67–68, 74
Thomas J. Watson Research Center, 164, 197, 203
Tsui, Daniel, 12
TTF-TCNQ, 97

U

Ulrich, Jochen, 144, 232

V

Van Dover, Bruce, 18–19, 66, 179, 226, 227
Van Parijs, Luk, 188–189
Varma, Chandra, 160, 176, 179, 193, 202, 205
Venema, Liesbeth, 125, 219–220
von Klitzing, Klaus, 92, 231, 232

W

Weber, Chris, 130–132
whistleblowing, 6, 183–184, 186, 190, 215, 218, 238

Whitesides, George, 166
Wigner crystal, 230
Wilk, Glen, 226
Willett, Bob, 168, 174, 203, 214, 216–218, 225
William L. McMillan Award, 158, 222, 227
Williams, Stanley, 164
Wilson, John, 161
Woelfing, Bernd, 81
Wudl, Fred, 78–79

Z

Zeis, Roswitha, 232
Zhitenev, Nikolai, 179, 202
Ziemelis, Karl, 63, 108–109, 119